RUTHLESS TIDE

ALSO BY AL ROKER

The Storm of the Century

RUTHLESS TIDE

THE HEROES AND VILLAINS OF THE
JOHNSTOWN FLOOD,
AMERICA'S ASTONISHING GILDED AGE
DISASTER

AL ROKER

WILLIAM MORROW
An Imprint of HarperCollins*Publishers*

RUTHLESS TIDE. Copyright © 2018 by Al Roker Entertainment. All rights reserved. Printed in the United States of America. No part of this book may be used or reproduced in any manner whatsoever without written permission except in the case of brief quotations embodied in critical articles and reviews. For information, address HarperCollins Publishers, 195 Broadway, New York, NY 10007.

HarperCollins books may be purchased for educational, business, or sales promotional use. For information, please email the Special Markets Department at SPsales@harpercollins.com.

FIRST EDITION

Designed by William Ruoto

Library of Congress Cataloging-in-Publication Data has been applied for.

ISBN 978-0-06-244551-3
ISBN 978-0-06-287593-8 (Barnes & Noble signed edition)
ISBN 978-0-06-287592-1 (Books-A-Million signed edition)

18 19 20 21 22   LSC   10 9 8 7 6 5 4 3 2 1

CONTENTS

CONTENTS

RUTHLESS TIDE

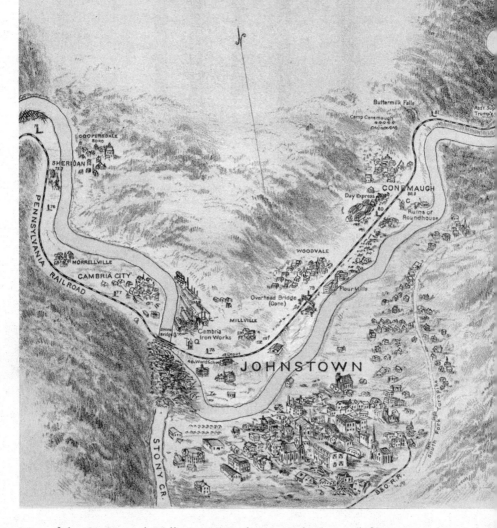

Map of the Conemaugh Valley in 1889, showing Johnstown (left) and the South Fork dam (right). CREDIT: Library of Congress

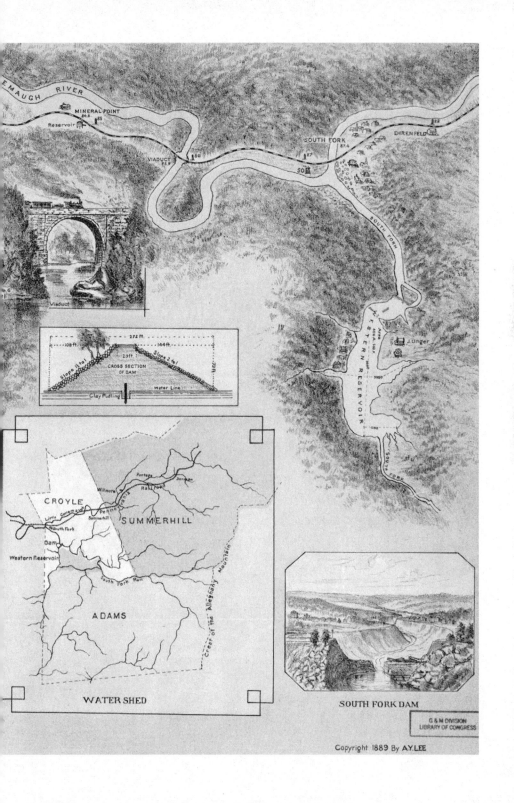

EMAUGH RIVER

MINERAL POINT
Reservoir

VIADUCT

Viaduct

SOUTH FORK

EHRENFELD

SOUTH FORK

Area 484.4 ft

J. Unger

WESTERN RESERVOIR

SOUTH FORK

272 ft.
108 ft. — 144 ft.
20 ft.
Slope 1½ to 1 — Slope 2 to 1
CROSS SECTION OF DAM
72 ft.
Water Line
Clay Pudling

CROYLE

Portage

Willmore
Railroad
Jamison

Little Conemaugh
Summerhill
South Fork

SUMMERHILL

Dam

Western Reservoir

South Fork Run

ADAMS

Crest of the Allegheny Mountain

WATER SHED

SOUTH FORK DAM

Copyright 1889 By A.Y. LEE

"MR. QUINN IS TOO FEARFUL"

Johnny, who made the world?" the Sunday school teacher asked.

That was easy. "The Cambria Iron Company!" the boy replied.

That's the story they liked to tell, anyway, in and around Johnstown, Pennsylvania, in 1889. Little Gertrude Quinn was only six that year, and even she, like the fictitious Johnny, knew how important iron and steel and the Cambria Works were to the life that she, her family, and everybody they knew were living here in the deep valley of the Conemaugh River under the abrupt rise of the Allegheny Mountains.

At about 3:00 P.M., on May 31, 1889, an explosion of previously unimaginable force would smash that world. Twenty million tons of towering water, released all at once from high above the town, falling with gathering momentum down the narrow Conemaugh Valley, picking up houses, factories, forests, railroad tracks, locomotives, livestock, and human beings as it came, would arrive with a roar like nothing heard in Johnstown before. In moments, and then in the horrifying hours that followed, that gigantic, raging thing would destroy thousands of lives, knock down and carry away hundreds of buildings, spread raging chemical fires, and send little Gertrude Quinn on a horrifying journey.

It would be the end of the world. And the Cambria company, which had made that world, would be powerless to stop the destruction.

Nobody in Johnstown could see what was coming, of course. They'd grown used to the world the Cambria company had made, a new world. At six, Gertrude Quinn wasn't old enough to remember that not so long ago, there hadn't been much in Johnstown but a sleepy farming village. But grown-ups could remember. That village had been succeeded by a canal town that seemed, for its time, comparatively bustling and excited, but by 1889, the former excitement seemed like nothing. The canal was defunct, replaced by railroad trains that chugged and screeched day and night along steel rails on the ridges: the foundries and steel mills on the riverbanks were going almost all the time, their fires sending up a perpetual haze of stink and emitting a steady pound and roar. Men drenched in sweat, their skins darkened by sooty smoke, moved quickly, athletically, on those factory floors, shoveling coal, carrying molten metal in long ladles, operating the huge, dangerous rolling machines that pushed out miles of steel rail.

Steel rail. That's what made not only Johnstown, Pennsylvania, but the whole nation run, and run in a new way. In a single generation, the United States had become a sprawling, booming power built on speed, noise, and engines, on electricity and steam, on iron, and especially on steel, cheap steel, hundreds of thousands of miles of rail carved by rolling mills from the soft, glowing product, cooled to an intense durability, and laid down for track that let engines haul hundreds of cars filled with still more steel, steel for bridges, for buildings, for a whole quickly rising America where dazzling fortunes were made and hard factory jobs were plentiful.

All that had begun here in western Pennsylvania. The Cambria company made not only the world that Gertrude Quinn and other Johnstowners knew but the new nation, too. By 1889, the company's vast rolling mill in Johnstown, with its many related factories and stores and yards, a virtual city of pounding, smoking, roaring work spreading along a tributary of the Conemaugh River known as the Little Conemaugh, was not only among the largest iron and steel operations in the world, but also the leading innovator in steel production. From the fictitious Johnny of the local joke to the real little girl named Gertrude Quinn, from the workers on the factory floors to the executives in the big houses to the bosses and investors living in glittering splendor in Pittsburgh—only an hour-long train ride west, these days—really everybody in western Pennsylvania knew that all kidding aside, Cambria Iron really did, in many ways, "make" the world they lived in, locally and nationally.

And steel spawned other enterprises. Entrepreneurs had come to the Allegheny valleys to found not only iron and steel factories but also the mills that added value to steel: the Gautier Wire Company, for one, in nearby Woodvale. Factories enabled by the steel boom made things as seemingly unrelated as woolens. Almost everything was tied up with the loud, relentless pace of the iron and steel economy, and in Johnstown the Cambria company owned most of those secondary factories, too. Iron and steel, so much of it made right here in western Pennsylvania, made America.

The most famous of the innovators who had become millionaires in railroad, telegraph, iron, and steel was Andrew Carnegie. Carnegie was changing the way all of "big business"—a new term in the 1880s—was conducted, and he was a man of western Pennsylvania. His own mills were located not in Johnstown but outside Pittsburgh, yet by 1889, innovations made by the author of *The Gospel of Wealth*, putting smart investment ahead of hands-on operation, as well as his astonishing adroitness in forming monopolies, had begun changing how things were done even by the all-important Cambria company. Carnegie loved the Allegheny Mountains and the Conemaugh Valley. He thought on planes higher than those represented by mere steel and business: he was interested in beauty, nature, contemplation, relaxation. Those interests, as put into practice on a mountain above Johnstown, were about to have an unintended, disastrous effect on the city and the entire Conemaugh Valley.

Little Gertrude Quinn's family wasn't in the steel business, not directly. She'd had a great-uncle who'd worked as a steelworker, but the Quinns neither labored as millworkers nor managed labor as executives nor counted their millions as investors. The Quinn family were store owners, and prosperous ones at that.

Dry goods: the business had been begun under the little girl's maternal grandfather, a Johnstowner of German extraction named Geis, now retired, when the town was far quieter. The store had thrived then, and Gertrude's father had taken it over, and with the boom in factory life and all of the big changes in town, the place had become quite upscale. The Quinns and Geises owned other properties now, too, right in the center of town.

By six, Gertrude had distinguished herself as an unusually sharp and observant little girl, full of life and curiosity, at times inclined to be naughty, and she loved the family store. Clerks and customers alike treated the Quinn children as important personages, and with few opportunities for organized entertainments, the little girl treated the store as her own personal show. It was brightly lit, and people not only came and went but also hung around, exchanging gossip. She especially liked a line of needles the store carried. They came in small bronze packages, on which were printed the faces of Gertrude's beloved Papa, Mr. James Quinn, and her uncle, a partner in the business. Gertrude adored Papa. Both men wore the perfectly trimmed Vandyke beards of the day: on the needle tins, the bronzeness seemed to her to make them look very fierce and strong.

Gertrude would steal these needle packages and proudly hand them out to her playmates. Her father didn't know that her friends' mothers were getting their Quinn store needles free.

Gertrude's mother was, to the little girl, always a smiling, protective, and devoted presence, but in fact Rosina Quinn was anything but the stereotyped "angel in the house" of Victorian ideal. Rosina, born Geis, was a smart and efficient person of business. In about 1859, her own father, having founded the company, took Rosina out of boarding school at the age of fourteen and set her up with a store of her own, right on Main Street, with an inventory of four thousand dollars' worth of merchandise. No woman in Johnstown had run a commercial establishment before, much less a teenage girl. With her sixteen-year-old brother, Billy, doing the books, Rosina's store took off. She was a tough dealmaker with a comprehensive grasp of all things commercial. The store flourished throughout the Civil War years, and Rosina didn't even give up the business with her marriage to James Quinn.

And yet when the Quinns' first child was born—Gertrude's brother Vincent—Rosina did step aside, and James Quinn joined the main Geis store. So while Rosina, like so many other women in 1889, of every class and background, now had the multitude of children that kept her frequently pregnant and endlessly busy—at six, Gertrude had three older and two younger siblings—and while Rosina loved, enjoyed, and took pride in her children and her home, she badly missed the business life.

Young Gertrude, though, wasn't especially interested in all that. Mama was just Mama, and she was everything anyone could want in a mother, as far as Gertrude was concerned. The younger children had a nurse to help out, but Rosina was as hands-on with motherhood as she'd been in business, tending to their needs and fears, wiping tears, reproving bad behavior, and teaching the children to pray not only in English but also in German.

Papa doted on Gertrude, but he was a dignified man, a believer in discipline, with his high collar and that neat Vandyke beard. Sometimes he called her his "little white head," for her white-blond hair. But at other times Gertrude had known the blows of a rolled-up newspaper, Papa's favored instrument of correction, on her bottom. She could never seem to remember to obey his commands until after she'd already disobeyed them. In one such encounter, her father brought down the rolled-up newspaper and felt it hit something strangely metallic. He hit again—the same thing. It emerged that the little girl, knowing she was in for it when Papa got home, had put a pie pan under her clothing for protection.

That was Gertrude. Luckily, Papa had a sense of humor.

It was Gertrude's older brother Vincent who had given her

that idea, and in return she had to run errands for him. He was sixteen. She adored Vincent, though she scrapped with him from an early age. All the Quinn children above the age of infancy had a lot of rambunctious fun in the rambling brick house at the corner of Main and Jackson streets, enclosed by an iron fence, one gate on Main, another on Jackson, and porches on each street, too. The house was fitted out with all the modern conveniences—a coal cook stove, central heat via a coal-fired boiler, indoor plumbing—that were Gertrude's mother's pride and joy, and the Quinns held big rounds of parties there. Thanks to the success of their store, the Quinns were known throughout town by the rich, middling, and poor alike, and they in turn rubbed elbows with some of the biggest people not just in Johnstown but throughout the region. They knew the Fritz brothers, for example, international figures in iron and steel. They'd known the late Daniel J. Morrell, the imposing partner in and general manager of the Cambria company itself.

Across a lawn on the same property lived Gertrude's maternal grandparents, who had started the family business, in a little white-painted brick house with green shutters. Fruit trees, vegetable and flower gardens, a barn for the milk cow Daisy, and a black-and-white dog named Trump rounded out the Quinn place. And Vincent, the firstborn Quinn, brilliant, funny, and already enterprising, kept pigeons and ducks.

But perhaps the best feature, from Gertrude's point of view, was the third-floor nursery, an entire story dedicated to the children's play. Fully fitted out by the Quinn store, this was a miniature home of its own, with a center hall, a parlor at one end, plush furniture, a nice carpet on the floor, little bedrooms for dolls—also fitted out with tiny beds, bureaus, and carpets—a dining room with a painted table, chairs, sideboard, dishes, tablecloths (hand-hemmed), and flatware. In the dining room stood a big wardrobe for storage, and across a small hall even a tiny kitchen, with a miniature iron stove and cooking utensils hanging from hooks on the wall. In a separate big, open space, Vincent and his friends roller-skated, while a castoff organ played songs. An amazing racket reverberated through the house. Somehow the Quinn parents put up with it.

Outside the house, Gertrude was always poking into things, alone and with her group of small friends. They were sometimes drawn to the Salvation Army hall, a new thing in town. The kids would stand outside the door, enraptured by the big drum and the women in bonnets hitting cymbals and shaking tambourines and singing.

"Oh! You must be a lover of the Lord . . . or you can't go to heaven when you die . . ." the Salvation Army people sang. The catchy tunes got in Gertrude's head.

Her main desire, at the age of six, was to be considered a true Republican. James Quinn was a staunch member of the Grand Old Party, having joined the Union Army at the age of twenty, right after President Lincoln, a Republican, called for seventy-five thousand volunteers; he'd served with the fancily turned out Johnstown Zouaves. With trinkets and candy Papa and Vincent had bought Gertrude's loyalty to the Republican cause.

She considered it the high point of her life, therefore, when Benjamin Harrison was elected president of the United States in November 1888. Johnstown held a torchlight parade, with band music and fireworks. Thousands of people filled the streets, and because of her politics, the girl was allowed to stay up late and sit beside Vincent on the side porch. Vincent was launching fireworks as happy, celebrating Johnstowners flowed by below. When he grew overwhelmed by the sheer thrill of it all, he would grab Gertrude by the skirts, lift her over his head, and shake her in triumph as if she were a rag doll. She loved it.

◆ ◆ ◆

At six, that is, Gertrude Quinn already had quite a personality. She was going to need it.

When it started raining on the night of Thursday, May 30, 1889, Gertrude's mother was visiting family in distant Scottsdale, Kansas, with even younger Quinn children. But Gertrude's aunt Abbie—her mama's sister—was staying with the Quinns, along with Abbie's own baby, Richard: Aunt Abbie had come east from Kansas in hopes of improving her health after the baby's birth. Gertrude and the other Quinn children were enjoying the visit, and things only got better when the rain came down.

By noon on Friday the thirty-first, streets were flooded, and water was already over the curbstone surrounding the Quinns' corner house. The house sat high: there were three or four steps up to a terraced lawn above the street, then more steps up to the porch, then a step into the house. Flooding was common in Johnstown, especially in the spring, and James Quinn had gone to the store to supervise the moving of his inventory to upper floors for safety. He'd told Aunt Abbie and the children's young German nurse, Libby, not to let the children out: he didn't want them getting wet feet and catching cold. And if the water got too high, he said, he'd be taking all the children up to the hillside.

But Vincent was already out splashing through the water, offering help to the merchants and friends preparing for the flooding. Everybody was getting ready: businesses and even most of the factories had shut down, as people waded in hip boots and raincoats, boated, and rode horses and dray wagons to move their valuables, and in some cases their families, to higher ground.

There was no chance that the bold and inquisitive Gertrude would resist the urge to get her feet in that water. Eluding Abbie and Libby, she got outside, and as the rain kept falling the water began inching in a fascinating way into the Quinns' yard. Soon it began covering the lawn like a strange and funny pond.

Gertrude sat there on the bottom step of the porch with her feet in the water, shoes and all. The water was yellowish. It gurgled. Vincent's ducklings were swimming around in the big yard, and Gertrude wanted to play with them. She kept reaching out for the ducks, her clothes getting soaking wet.

Suddenly she found herself jerked to her feet. Libby, having caught a glimpse of the girl playing in the water, had run outside, and was now dragging her into the house to get dry and change her clothes.

Not long after noon, James Quinn came home as usual for dinner. Unlike most of the other people out preparing for the flood, he was worried and restless. James wanted to get the whole family up to the hills before the water got too high.

But Gertrude's baby sister Marie had the measles. While the child seemed to be recovering, she was still weak. Her father was loath to get her out of a warm bed and darkened room and into the rain and the daylight: light, it was feared, could harm infant measles patients' eyesight. Cold rain might cause a relapse.

So at dinner, Mr. Quinn issued new orders to Abbie and Libby: keep the children ready for a run to the hillside. Should the rain get bad enough and the water high enough to risk the baby's health by making such an escape, that's what they were going to do.

James Quinn had a reputation for worry. His anxieties had long since attached themselves to a single, even an obsessive subject. People around town found it at once somewhat irritating and somewhat comical. James was forever worrying aloud that if a bad enough rain came, flooding would be the least of Johnstown's problems.

Now, at dinner, he said it again.

If the dam should give way, he reminded Aunt Abbie, not a brick would be left standing in Johnstown.

To Aunt Abbie, her brother-in-law's concern seemed extreme, even silly. This flooding was unusually high, it was true, but flooding was common. She laughed and reminded him how strong his big house was, and when he went back to the store, Abby laughed again.

"Mr. Quinn is too fearful," she told nurse Libby.

The dam that James Quinn was always fretting about, and with special restlessness on that rainy last day of May, loomed like a titanic, ancient earthwork to the northeast of Johnstown, up the valley of the Little Conemaugh River, not only fourteen bending miles upstream but also nearly five hundred feet uphill. Like many valleys in the Allegheny Mountains, the Conemaugh tilts steeply, its tight twists and bends bringing runoff down fast from a multitude of tributaries. At the bottom of that drop, Johnstown's homes and factories and stores and public buildings stood on an unusually flat point, formed by the confluence of the Little Conemaugh and the Stony Creek—headwaters of the Conemaugh proper—a city in a kind of geological hole in the steep, forested rock that rose around it on all sides.

The dam up the valley had been completed back in 1881. For eight years now, it had existed over Johnstown, so high above yet so near at hand. More than 60 feet high and 900 feet long, a curving form 270 feet wide at its base but only 20 at its flat top, with a roadway across the top from hillside to hillside, the dam offered a fantastic view out over the whole valley, and for all of these eight years it had been holding back more than 400 million cubic feet of water, about 20 million tons, in a man-made lake more than two miles long and about sixty feet deep. The lake was the creation—the pride and joy, really—of an organization called the South Fork Fishing and Hunting Club.

On the afternoon of May 31, 1889, just as James Quinn had feared, the dam broke, and 20 million tons of water started heading down the valley. The valley was fully populated, and, well before it arrived in Johnstown itself, that vast destructive force had taken out whole towns, factories, most of the human endeavor of the Little Conemaugh Valley, along with entire swaths of forest and grass. Not much remained in the path of that incredible wave after it had poured, a charging, churning wall full of gigantic objects, down the valley, stripping land right down to bedrock. The Great Johnstown Flood of 1889 would live long in the ranks of American natural disasters.

But like many other so-called "natural" disasters—including the worst ever in the United States, the Great Galveston Hurricane of 1900—the Johnstown Flood became such a gigantic human tragedy not fundamentally because of the terrible violence that can arise from turbulent relationships among climatic and geological and atmospheric forces. Those turbulences played their parts. But the 1889 Johnstown Flood was, more crucially, an unnatural disaster. The destruction it left behind was a human creation, one reflecting some of the least happy tendencies of human beings.

It's true that many of the flood's contributing factors involved nature at its most threatening to human life. Rain fell, for example, for weeks in torrents never before seen. As far from Johnstown as the wide streets of Washington, D.C., in the first week of June people were wading in water two and three feet deep. The flooding during those days came from the usual source: buckets and sheets of pounding rain, with the rivers' concomitant swelling past the point of any potential for normal drainage. Nature caused a lot of damage at the end of May 1889.

But even the nearly annual flooding that Johnstowners had gotten used to wasn't *naturally* normal. It had become so regular that people like James Quinn's sister-in-law Abbie, along with so many others in town, laughed at those who feared a disaster would one day strike, but the frequent appearance of high water was new, and its most immediate cause lay in erosion, caused by massive lumbering in the forests on the hills above the city. Timber was in demand for housing to support and fulfill the employment boom that had come to mark life in the valley. Industry itself had brought quick, extreme ecological change to the region.

And yet industry, people believed, was not only inevitable these days but also generally a great thing. True, endless rows of stacks issued black smoke day and night, laying a haze of stink along the valleys of the Conemaugh and the Stony Creek, often putting a thick, smelly cloud over Johnstown. Many days you couldn't see sky. True, too, birds no longer sang here, the trees near the rivers had all turned black and never leafed anymore, and fish no longer jumped in those rivers where they flowed past the city. Add the scarring effect of the cutting on the hillsides, where bareness above the factories contrasted sharply with the green of the upper ranges. And through all of this noisy, burning, blackening, pounding, scarring, stinking activity came the locomotives, huffing and screeching on the steel rails, long freight trains pulled by steam and stoked by coal, shaking the whole ground.

But all of that only reflected the exciting fact that two-thirds of all American steel was now made right here in the river valleys of western Pennsylvania. Industry meant work. The population of Greater Johnstown alone had doubled from 15,000 in 1880 to 30,000 in 1889. Descendants of the first European settlers, including early-arriving Germans, had been joined in large numbers by new immigrants from Italy, Poland, Hungary, Bohemia, and elsewhere, all seeking work. It was labor, and the management of labor for the maximum profit of investors, that drove this production.

Those people had to buy not only food but also things like dry goods and notions. Well beyond the steel industry itself, in places like the Quinn family's quietly humming store, the money and management and labor that went into steel, and turned the small, flat plain at the bottom of the valley into a cacophonous inferno, offered a backbreaking subsistence for many, upward mobility for others, prosperity for some, and fabulous wealth for the bosses. Most of the people in Johnstown, glancing up at their partially bare hillsides, didn't wax nostalgic for a simpler, quieter, and more peaceful time. They were busy. Pretty sights weren't necessarily on their minds.

Trees, however, aren't just pretty. They hold water. Removing them in huge swatches means that in heavy rains, water once drawn upward by the vascular action of the xylem and phloem of big trunks, enabling photosynthesis to amplify a flourishing canopy, has nowhere to go but down. So more water flowed now, into the tributaries and bigger rivers that drained the Alleghenies. Since the rivers couldn't handle that much water, they overflowed instead, and they spilled over the most, of course, across the floodplain at the bottom, down in the Johnstown hole, where there was room to spread, fill basements, and turn the streets of Johnstown into rivers.

Other unnatural facts amplified the effects of the excess water now coming down the hills. Even as more water flowed into them, the rivers were being narrowed drastically. The factories' slag and other massive, sludgy waste—ash, used coal, molten dross—were poured and thrown, constantly and without any thought, right into the Conemaugh and the Stony Creek, detritus of the unceasing glow of orange coal fires and the smelting of iron to steel. A hot sludge poured freely into the flow, then cooled into hard, junk metal underwater. Over time—and not much time at that—waste blocked up the rivers.

Narrowing had another cause as well: the Cambria company was developing along the riverbanks, using landfill to build out new space near the water to accommodate worker housing and railroad beds and other construction convenient to its great works and yards.

So between erosion from lumbering on the hills, the narrowing of the channels by waste, and landfill for construction, more and more water had to be drained, especially in the spring, in less and less space. The very industries that were attracting so many people to live here, in such new droves, and had quickly converted the whole Conemaugh Valley into a sprawling, always-on industrial enterprise, also contributed to the flooding that would take so many of those same people's lives at the end of May 1889, traumatize their survivors, and snatch children from parents and parents from children and spouses and friends and relatives from one another in a turbulent horror. The investment in industrial manufacturing that was making America a commercial empire for the first time—and creating, for some, the standard of living justly celebrated by little Gertrude Quinn's house-proud mother—gave the people of Johnstown the work by which most of them eked out hard livings and others prospered nicely. That industrial investment had so quickly altered the natural conditions of the valley that when it started raining at the very end of May, many people in Johnstown at first took the day's deep flooding for granted.

◆ ◆ ◆

Yet even those human alterations of nature, with so many unintended consequences, don't give us the ultimate cause of the sudden, horrific loss of life, and the stunning loss of so much else both natural and human-made, that occurred so fast when the towering monster that became the Great Johnstown Flood was unleashed.

James Quinn was right. Nothing like that horror would have occurred had the dam at the South Fork Creek, a tributary of the Little Conemaugh, not let go, all at once, the entire tonnage of artificial lake it had been holding back. Unusually heavy rains, erosion, the narrowing of rivers: they played their parts. So did certain unusual atmospheric conditions that held the rainstorm over the Alleghenies for so long. But the release of the lake caused the disaster.

Had the dam up on the mountain not broken, Johnstown would have seen record high water that day, and that's about all Johnstown would have seen. There would have been damage. One life was lost to flooding, that day, before the dam broke. Conceivably there would have been a few more deaths.

The dam's breakage caused a disaster incalculably greater than that. It was, some said, an "act of God." Certainly that's what the owners of the dam and the lake would say, when it was all over: what dam wouldn't have succumbed to such epic rainfall, such record flooding? The South Fork dam may not have been natural, yet surely the breakage was a result of nature at its most violent.

But nothing in nature determined that the dam had to break that day. Human decision-making issued that decree. The fabulously wealthy captains of those same industries that were changing the face of not just western Pennsylvania but the whole United States had perfect faith in their own imaginations and in their own collective judgment. They'd enjoyed extraordinary successes to date. They'd dreamed up projects nobody else could have imagined, and they made those dreams into reality. They believed they could accomplish anything they wanted and that anything they didn't want to happen couldn't happen. Up on the mountain, they'd dammed the South Fork and let nature take its course, and a big, beautiful lake had appeared. The members of the club didn't want their dam to break, of course. That was the last thing they wanted. They loved their lake.

They could have kept the dam from failing, but because they didn't want it to fail, they believed it couldn't fail and they took no steps to keep it from breaking. It broke anyway, and the lives of thousands of people were lost in agony, and the lives of their survivors were thrown into another kind of agony, because those men weren't accountable to any power on earth.

PART I

MEMBERS AND NONMEMBERS

CHAPTER 1

UP ON THE MOUNTAIN

In the spring of 1889, when an event whose only comparisons were biblical descriptions of the awful Last Day of Judgment came rushing into Johnstown, few people in the valley knew for certain who belonged to the South Fork Fishing and Hunting Club, the private retreat up on the mountain, with its marvelous, sparkling artificial lake. Almost all of the club's members lived in Pittsburgh, not Johnstown, and they weren't the kind of men who wanted their private affairs bandied about for all to see.

Some were already world-famous and didn't need gawkers coming around. Many weren't famous, but they too liked the privacy, and maybe even more they liked the sensation of belonging to a club with men who did have such needs. These were steel magnates, railroad tycoons, coal barons, ironmasters, glass producers, and other entrepreneurs in the process of creating what Mark Twain was soon to call the Gilded Age, along with the lawyers and bankers and insurance men who served various processes for the great industries that those men were building.

They all did business together, less in cutthroat competition than in a strange kind of communalism, sewing up business in impromptu, often undocumented consortiums, seeking and often achieving industrial monopolies.

For it was in monopoly that the secret of great American wealth lay. These few men, centered in the booming city of Pittsburgh, created and held a huge piece of that wealth. In its pursuit, they invested in wresting ore from the mountains and altering molecular structures. They conceived and then actually brought into being great networks of water, rail, human toil, and financial capital, to spread the products of their desire and ingenuity from coast to coast. These efforts, with the rewards they birthed, had become the signal American achievement, celebrated not only in the United States but around the world as harbingers of a coming new century.

If such men could conceive whole new industries, new technologies, and new ways of moving capital around the globe, surely they could manifest for themselves a rustic retreat, far above all the haze and smell and noise that their imaginations were creating down below. Up where the air was still cool and clear, they could imagine big bass gobbling hooks and fighting lines, wildlife succumbing to bullet and shot, well-cared-for youngsters darting gaily colored sails about a mountain lake, and summer cottages—meaning oversize "gingerbread homes"— with docks and boathouses along a lakeshore. Whatever they could imagine they could make real.

What they couldn't imagine, and what none of them would ever consider, let alone admit, was that they would be the cause of the greatest disaster to that point in American history. Catastrophe was so far from their purpose, it scarcely entered their minds.

The members had two goals in forming the South Fork

Fishing and Hunting Club. They would seek relaxation and recreation in healthful outdoor sport, and they would provide their growing families a place to cavort, in a genteel way, with others of their station. In those pursuits, the members of the club expected to remain undisturbed, beholden to nobody not in the club. They formed the organization in 1879 and had it going in 1881. The membership rolls remained secret.

In 1889, the most famous name to be found among the secret membership rolls of the South Fork Fishing and Hunting Club was Andrew Carnegie. Nobody in the United States had more money, it was said; some counted him the richest man in the world. And Carnegie gave his money away. Libraries, concert halls, and other great institutions of culture and learning were beginning to bear his name from coast to coast.

And it was Andrew Carnegie, of all the South Fork Club members, who most fully expressed the new spirit of the age. That spirit had changed the Conemaugh Valley, in a few short years, from a farming town to a canal town to the archetype of the loud, harsh, smoky, clanking, polluted industrial environment, stripped of trees. From the workers' mere survival, in the hard labor of the mills and mines, to the genteel classes' prosperity, in the stores and offices, that spirit presaged the dawn of modern America, and Andrew Carnegie was becoming world-famous for leading the effort.

But what Carnegie most epitomized was the interconnectedness of all industry via capital. By 1889, he'd made deals that linked his own steel company with that of the Cambria company of Johnstown, which little Johnny thought made the world. Carnegie competed with, poached from, and appropriated many of the innovations in production that made Cambria great; and by

1889, for all of Cambria's might, he was already beating it, in part dictating to it. And Carnegie didn't only own steel mills— soon almost all of them—in western Pennsylvania, where so much of the world's steel was produced. He owned or held powerful interests in the coal that fired the steel hearths, and in the railroad trains that hauled the coal to the mills and hauled the steel products away, and even in the hotels where his fellow businessmen stayed when pursuing their many great aims. Working closely with those who might otherwise have been competitors, Carnegie's consortiums were financed by banks in which he also held powerful interests, allowing him to "capture markets," as he liked to put his ceaseless quest for monopoly.

To the Pittsburgh industrialists and businessmen, Andrew Carnegie was thus not only their partner and their leader but also the model to which they all aspired. And not only in business.

When the rich men of Pittsburgh formed the South Fork Club up on the mountain, Andrew Carnegie was their guiding spirit in that effort, too. While his rise to fabulous wealth and fame involved an extraordinary degree of sheer toughness, more important to the founding of the South Fork Club was another side of Carnegie: a love of contemplation, artistic beauty, philosophical thought, and natural wonder.

It was that combination of qualities, seemingly incongruous to later generations, that made Carnegie the highly idiosyncratic American icon of wealth and philanthropy that he would become. That combination formed the mentality that enabled the men of the idyllic mountain club to revel in their pastoral, romantic mountain retreat to the exclusion of any sense of the danger they were posing thousands of people who were living and working below. Often remembered later as the hard-nosed if generously philanthropic capitalist archetype of the booming America of the late nineteenth century, Andrew Carnegie was really a romantic. It was he who first started promoting the healthful benefits that might accrue, to a hard-driving, urban, modern man of business, from the physically and mentally restorative powers of recreation in a natural, breezy, pastoral, mountainous retreat. That idea, in such stark contrast to the scarring of the countryside and the fouling of the air and the blocking of the rivers that Carnegie's and others' industries were causing throughout the region, inspired the members of the South Fork Club to build a mountain retreat. His colleagues were copying Carnegie's peripatetic search for rest and recreation in beautiful mountain settings.

His joining the South Fork Club gave them their apotheosis. Why they dammed the river and made a lake, and why they refused, in the face of multiple warnings and expert advice, to prevent the dam's breaking, and to prevent the lake's becoming a biblical monster of destruction, can only be understood—or approached, anyway—in the irresistible rise of Andrew Carnegie, the relationships that formed through and around him, and the meaning to Carnegie and his fellow industrialists of the joys of mental restoration in a simple, natural setting.

That rise to extraordinary wealth and power wasn't a fake tale. Andrew Carnegie really did begin life in poverty and really did become one of the richest men in America. He truly did live and help invent a vaulting part of the American Dream.

His biography would thus become perhaps one of the best-known rags-to-riches stories of the age. For well more than a century after Carnegie's birth, ambitious boys and young men would pore over cheap, repetitive editions of his life in search of inspiration and encouragement. The career seemed to give proof to the notion that in America, any lad possessed of the right virtues could grow up to become fabulously successful, rich beyond the wildest dreams of avarice, and yet immortal for goodness and benevolence, with his name on every philanthropic effort you could see.

The story remains pretty awe-inspiring. Andrew Carnegie was born and raised far from Pennsylvania and, unlike most other members of the class he would ascend to, he had no role models to guide his ambition. He came not from wealth in the United States but from poverty in Scotland. The amazing transformation in Carnegie's life encapsulates both how the whole region around Johnstown could be changed so utterly and quickly, and how it could then be so devastated by death and destruction.

He was born in 1835, in Dunfermline, long a center for making textiles, when the famous cottage-industry approach to textile production was well under way. In that system, workers were saddled with some of the worst aspects of both wage slavery and tenant dependency: living conditions for the poorest laborers had become worse, in certain ways, than for some of the medieval peasants they'd superseded. Andrew Carnegie, called Andy by his family, was born in a two-family house, one room for each family, typical of weavers' cottages, and better described as a semidetached brick hut. His father, William Carnegie, worked a hand loom there, making linen damask for fancy drapes, and at first the Carnegies seemed to be moving up in the world. The family soon moved to a larger house.

Then the work disappeared overnight. The whole region fell into a state of literal starvation.

For "progress" had occurred. Industrialization had developed. The boy who would grow up to become one of the richest industrialists in the world saw things go, nearly overnight, from bad to outright impossible when the steam-powered loom, newly perfected, came to Dunfermline and changed everything. Andy's father went from being a skilled laborer, moving up somewhat in the world, to being a displaced and unemployed man begging for work, deeply humiliated and disgraced. Just like that.

Automation actually came late to Dunfermline. The trend had been under way for some time in the British Isles' textile industry. With sudden shifts to factory work, skilled laborers were being thrown out of work. Children could now be set to work for long hours. Injuries became rampant. In some parts of England, skilled workers rioted in protest over this wholesale, overnight reorganization of labor, by which they felt they lost not only work but ancient rights: Luddites, some of them were called. Rallying behind a fictitious figure they sometimes called Ned Ludd, after an actual rioter, and other times called King Ludd, they attacked and burned factories that were using the new stocking frames, spinning frames, and power looms.

All that had begun early in the century, and the Luddites' labor actions had been put down with force by the British Army. But the power loom didn't get to Dunfermline until the 1840s, when young Andrew Carnegie was old enough to see and feel its effects on the already overwhelmingly harsh lot of postfeudal laborers.

Andy might have become a kind of Luddite. He could have viewed technological and industrial change as a deadly force. In the end, he would go the other way. Instead of resisting industrial change, he would embrace it in order at first to own it, and then to rise above it.

Amid the devastation brought on by the arrival of the steam loom, Andy's mother, Margaret, soon became the family's sole support. She went to work for her brother, a shoemaker, and she sold baked goods. And it was Margaret Carnegie who decided that the region around Pittsburgh, in western Pennsylvania, so far across the ocean, was the right destination for the desperate Carnegie family.

Here was wholesale, overnight change of a different kind. Andrew Carnegie would always have a healthy respect for the stunning speed at which things can transform, and yet far from fearing such forces, he would always hope to turn speed to his advantage. And even as a boy he would have known that a move to Pittsburgh probably made sense.

Like Dunfermline, in the 1840s and '50s Pittsburgh and its nearby towns were among other things a textile center. In America, cotton had recently become king, and the town of Allegheny, on the north side of the river of the same name, right across from Pittsburgh, had become a cotton brand. MADE IN ALLEGHENY: so read the labels on bolts of cotton sold throughout the country. Many other products bore that label too, from iron to rope, from wool to paper, from steam engines to linseed oil. Allegheny was putting people to work, and Margaret Carnegie had two sisters living there who testified to this. It was a tough town, with workers housed in shabby, crowded houses blackened by coal smoke. Yet all that dirt and smell came from human toil at steam-driven machines. To eat, there had to be work, and Allegheny was about labor and little else. Not long after the family arrived in Allegheny, William Carnegie was fully employed in a Pittsburgh cotton mill.

And so was Andy. Twelve hours a day and six days a week, beginning before dawn and ending in the dark, the thirteen-year-old changed out spools of thread for $1.20 per week. The job was at once boring and dangerous. The repetition was mind-numbing, yet kids were expected to reach adeptly into huge machines whose sharp needles were mercilessly driven by the hard force of steam, and fix any broken threads and parts. To say there were no safety regulations is putting it too mildly: nobody had ever heard of them.

At his second job, making a little more money, Andy had to run the steam engine itself. Down in the factory's cellar, his face and clothing thickly covered by coal dust, he would fire the boiler and work the gauges to send up enough steam to give the workers power to operate the machines, yet not so much that he would burst the boiler and be killed in the explosion. Things like that happened.

He also had to bathe new spools in giant vats of oil. That task literally made him vomit.

And yet Andy's daily life was typical for children in the booming industrial economy of the American Midwest of the mid-nineteenth century. The hard-bitten kids of Allegheny grew up tough, accustomed to discomfort and fatigue, and Andy did, too. Not typical was his rise.

Back in Scotland, he'd attended the free school, and his handwriting wasn't bad. That was an advantage not every boy had. He found himself occasionally taken from his harder labors to write up bills as a clerk. That work was easier, safer, more restful. Internally, too, Andy had some distinct advantages. He was filled with dreams that drew his imagination upward, far from the world of hard physical work. Being steeped in the literary and heroic wasn't necessarily regarded as an asset in the immigrant community of hard laborers in which he now lived, yet having been given at a young age a high sense of Scottish tradition and legend, Andy idolized figures like Robert the Bruce and William Wallace, men who led the clans against England for Scottish independence. Even as he worked these dull and deadly jobs, Andy Carnegie was mentally reciting the poetry of the great Scottish poet Robert Burns.

So as he labored in boredom and danger, he imagined following in the footsteps of great men, leading his family out of poverty and making a name that would resound in legend for all time. His voice still had the brogue. Some of the other boys taunted him as "Scotchie." Andy refused to take being Scottish as an insult.

Inveterate practicality played a part, too. Having learned to write up a neat single ledger, Andrew enrolled in a winter night school to learn double-entry bookkeeping. Soon he had a job lacking the physical rigors and dangers of the factories: messenger, at $2.50 per week, for the Atlantic & Ohio Telegraph Company office in Pittsburgh. It was like being paid to get out of prison.

That was 1850. Only five years later, Andrew Carnegie made his first investment.

That's how fast he worked his way up after leaving the telegraph company—they'd begged him to stay—and going to work for the Pennsylvania Railroad. Coming on as a private secretary and telegraph operator—he'd taught himself Morse code—Carnegie worked directly for the head of the railroad's Pittsburgh division, Tom Scott, and such was his eagerness and competence that he quickly drew the attention not only of Scott but also of J. Edgar Thomson, the company's president. They became his mentors.

It was in such relationships that Carnegie would thrive. And no matter what else he did later, he would always maintain valuable business connections with the Pennsylvania Railroad. That company was becoming the most important rail line in the world when Carnegie joined it, a multifaceted new blend of physical infrastructure, high-powered technology, and complex, high-stakes, high-speed daily administration. Undaunted by the new challenges of moving freight about on such a massive and complicated scale, Andrew Carnegie, a true company man, rose fast.

In 1855, Tom Scott invested $500 for Andy in an express company that managed the shipping of documents. It was a sure thing: the railroad company was giving the shipping company an inside-track deal, and Scott was taking care of young Andy by personally lending him the money for his first investment, thus welcoming him into the real processes of getting rich. This insider move, tiny and insignificant by Carnegie's later standards, exemplified the way all of his future business would work.

He received ten dollars a month as a dividend. He saw now that money, not work, is what made money grow, and that knowing who was who offered the key to investing money well.

Meanwhile, he was learning that cutting costs offered another way of making profits. Working closely with Scott, he began finding efficiencies throughout the Pennsylvania Railroad, and by 1860 he was the railroad's western division supervisor himself. It was clear to his superiors and to his rising peers that Andy Carnegie could make things work.

Government helped, too. During the Civil War, Carnegie was appointed by Thomson—once Carnegie's biggest boss at the railroad company, now assistant secretary of war for military transportation—to superintend all military rail and telegraph. Here was a movement of matériel and troops at a scale never before known in the United States. Previous gigantic movements of troops, like Napoleon's in Europe earlier that century, hadn't had the railroad. There was the quick laying of track, the massive scheduling, the demand for total reliability, the management of a newly critical volume of information communicated by telegraph wire, without which rail couldn't operate at all, at this scale. Carnegie executed the position with immense efficiency, and what he achieved in wartime presaged efficiencies and scales to come in the postwar world of industrial development. He came out of the war not only working for the Pennsylvania Railroad but as perhaps its most important executive. And now he had some real capital to work with.

In 1864 came Pennsylvania's oil boom. Carnegie put $40,000 into an oil-producing property, and thanks to the insane noise and bustle of hundreds of pumping derricks and teams of horses and barrel makers, with raw petroleum slopping and flowing everywhere—and thanks, Carnegie also thought, to the heroic, go-for-broke attitude of drillers in search of oil wealth—he drew a million dollars in cash dividends out of the property, without ever touching any oil.

Now he could go into the biggest business of all. In the 1860s, that was iron.

It took iron to make the locomotive engines and the tracks those engines ran on. On iron rails, train cars carried the raw materials of iron, mined from the earth, and the charcoal and coke that could turn the ore into product. Carnegie had his eye on becoming what was called an ironmaster. Big iron, and then the radical improvement of iron called steel, was to change the face of western Pennsylvania almost overnight, bring about whole new ways of working, attract laborers by the tens of thousands, make millions, and heat and smell things up so badly that the men who brought about that change would need to seek relief by the shores of a lake in the fresh mountain air.

Possibly about four billion years before the rise of Andrew Carnegie, radioactivity heated some of the asteroids in the belt between Mars and Jupiter to extreme temperatures. That heat separated the asteroids' melted rocky material from the melted metal at the cores, like oil and water: iron and nickel were isolated. They cooled within the asteroids over millions of years and then suddenly, with some collision or other catastrophic event, scattered as iron-nickel fragments.

Some few of those scattered meteorites came flying through the earth's atmosphere, melting again during that plunge. Then they cooled on land as hard iron and nickel rocks. There isn't much of this meteoric iron on earth, and when people began finding it, only maybe 3,500 years before the rise of Andrew Carnegie, they prized it for strength in weaponry and jewelry, as well as for sheer rarity.

But there was also stuff in the earth's crust that could be turned into iron. The element was obviously impressively strong, and ways of getting the ore out of the ground, and then getting the iron out of the ore—that rocky, non-useful stuff that iron is mixed with—began early, too. The ancient world mined for iron ore, and experts knew how to separate the molten iron from what would later be called the "slag," or everything not iron. The oldest iron furnaces were ingenious earthen or stone chimneys, packed with glowing charcoal. Hot iron and slag were pulled from the bottom of the chimney and separated, and sometimes small amounts of slag were mixed with the iron. Before the iron cooled, it could be hammered, or wrought, into a multitude of products, hardening into something amazingly durable, heated red hot again, reshaped, and so on.

Such was the technological inventiveness of people living thousands of years before the dawn of the American age of industry. Iron smelting went on in sub-Saharan Africa, in Europe, and in China, where the blast furnace first took over from the earlier chimneys and created what would become known as pig iron. That product emerged in fully liquid form and could be either cooled to create workable iron, or poured wet into molds to create what would be called cast-iron objects, fully set in complicated shapes, useful in days to come for ambitious projects like bridge building.

Such was the business Andrew Carnegie now had the capital to invest in. By the early eighteenth century, iron and other industries had already stripped the country of the forests for charcoal, and Abraham Darby began using coke instead of charcoal to fire his blast furnace in Shropshire, England. Coal mining to supply ironmaking began ripping countrysides apart and filling the air with coal ash. That was more or less the process that Andrew Carnegie would take up in the 1870s.

Soon Carnegie had opened three iron and iron-related companies in Pittsburgh: the Superior Rail Mill and Blast Furnaces, both a furnace and a rolling mill; the Pittsburgh Locomotive Works, turning iron into railroad engines; and the Keystone Bridge Company, which would use iron for the all-important spans that now had to hold up iron rails, locomotives, and hundreds of freight cars, passing back and forth day after day to make all this industry happen. When Carnegie and his partners proposed to span the Ohio River at Steubenville with a bridge made of both cast and wrought iron, people couldn't believe the iron would stand up, let alone support heavy train traffic. Iron bridges had failed before, leading to some of the worst railway disasters. But Carnegie had hired the best engineers and builders—hiring, more than ironmaking and bridge building, was among his key skills—and Keystone Works bridges were soon in use in nearly every state of the Union. In 1868, Keystone would span the Mississippi River at Dubuque, Iowa. That first bridge at Steubenville, despite all the initial skepticism, would continue to carry train traffic well into the twentieth century.

◆　◆　◆

Through all of that accomplishment, what Andrew Carnegie had shown himself especially good at was boldly embracing change and turning it to his advantage. The steam-powered loom had nearly destroyed his family. The brilliance of his investing, always via insider connections, had to do with seeing change coming before others did and owning it early.

His abilities also had to do with scale. Big, national ventures like rail, along with the telegraph, which enabled rail to run more or less reliably and safely, benefited from state and federal contracts, always tending toward the monopoly that Carnegie was relentlessly seeking. Such ventures boomed not locally, like oil or coal, but on broad scales; oil, coal, agriculture, manufacture, and every other form of enterprise were coming to depend on them.

Carnegie saw the future in another way, too. He learned quickly that the trick to making the easiest money wasn't by investing in a new rail company, for example, but by knowing where the existing rail companies were planning to lay track. He already had the connections for that: he sold iron rail to the Pennsylvania Railroad, where he still also served as an executive, and he knew the right people in other rail companies. They were consolidating fast and forming huge enterprises, and Carnegie was in a uniquely good position to learn in advance where their trains would run. This path to wealth involved efforts no more complicated than buying up land at low prices in far-flung places like Kansas, when nobody there knew how valuable that land was about to become.

Carnegie also formed a telegraph company, but not the kind that owned telegraph stations and hardware and sent and received messages. This business owned but one asset: an exclusive right—gained, again, via Carnegie's friends at the Pennsylvania

Railroad—to run wires along the railroad's existing telegraph poles. Without running a single line, Carnegie then sold that company, along with a contract to run wires from Pittsburgh to Philadelphia, combined with a right to run wires from St. Louis to Indianapolis, for a nice stake in the Western Union telegraph company, an investment that was to profit him enormously over the decades.

Nobody can say Carnegie wasn't in the telegraph business. Yet he'd never run a single wire, and by 1865 he had no time to run a railway company, either.

He had to tend to his investments. He was only thirty. Steel was still a few years away.

So Andrew Carnegie might easily have been taken, even at that relatively young age, for the ultimate hard-driving, all-business industrialist, a man with nothing but steam and oil and coal and rail and speed and stocks and bonds on his brain, a crude man of the booming factory Midwest, embodying entrepreneurial Pittsburgh, even though he stopped living there in the mid-1860s. Over the next twenty years, as the steel business began and then boomed, and as he became in monopolizing that business not only Pittsburgh's but also the nation's and then maybe the world's richest man, he naturally kept a big foot in his city at the Ohio headwaters. His frequent visits there were met with excitement and headlines and galas.

But where the old ironmasters knew iron, Carnegie understood investment, cost-cutting, vertical integration, and monopoly, and he knew how to hire managers. Because his real business was investing, selling securities in his ventures, capturing markets, registering patents, and directing subordinates not so much in hands-on management as in cost-cutting initiatives, he could live anywhere. And so he did. He traveled often to his beloved native Scotland, and he became headquartered in New York City. His mother, Margaret, who had brought the family from Pittsburgh, lived with Andrew in New York, in hotels: she liked having a staff, amenities, and a built-in social life, and she didn't want to manage a big household. Early on, and despite the work ethic that had played a role in his rise, Andrew Carnegie was working smart, not hard. Where his fellow Pittsburgh industrialists found it difficult to tear themselves away from the minute-to-minute business squabbles and stresses and gossip, Carnegie was having fun and improving his mind.

In the mornings in New York, he sent letters to his many managers. Then he visited art galleries, read books, and hosted big parties full of the most famous and interesting people he could find. They flocked to him, and much of his life now involved jovial, charming, and informative conversation. He wanted to write his own books, too, and further public appreciation of the arts. He wanted to convert people to his way of seeing the world, and of seeing money, and he wanted to improve the American mind.

He and Margaret did also live in Pennsylvania—just not in Pittsburgh. In the 1870s, Carnegie began making annual summer escapes from city life, up on the Alleghenies. He and his mother would travel to Cresson Springs and live in a house Andrew had built there, near the famous Mountain House, a resort hotel operated by the same railway company whose management he had given up. There, two thousand feet above sea level, at the crest of the mountains, the air was clear and cool in summer; far from the cares of business, and far from the excitements of glittering New York, the mind could come to rest on higher things.

For Carnegie, there was no contradiction between this pursuit of inspiring natural beauty, healthy exercise, and a religious sense of the sublime on the one hand, and relentless pursuit of business on the other. The book he wrote was *The Gospel of Wealth*, and he meant "gospel" not as metaphor but as reality. For him, the pursuit of wealth, in its most monopolistic form, with strong government backing, was tantamount to the good news of spiritual redemption for individuals—some of them, anyway—and nations as a whole. An inveterate romanticism fed both his remarkable business efforts and his desire to get away from it all. This ultimate man of business, with offices in New York and multiple investments in Pittsburgh, was still immersed in Robert Burns, still sure there was something bigger and better beyond industrial pursuits. The Alleghenies reminded him of Scotland—not the textile-town Scotland he'd actually known but the Scotland of Burns's burr and those brave tales of Wallace and the Bruce.

Dominating business, Carnegie didn't need to track its every move. A new idea was out there, and it wasn't Carnegie's alone. The simple life was the better life, both spiritually and physically. Mental restoration could be found only in return to nature.

Simplicity, but with amenities. After a few visits, Carnegie built a Queen Anne–style house up at Cresson. The nearby Mountain House, social hub of his world on the mountain, was a classic of the mountain retreat becoming popular among the rich and upper-middle classes of the day. By 1881, a new structure had been completed there, four stories tall with a front facade that spanned more than 300 feet. It could accommodate 700 guests, and an amazingly large dining room could seat 800 diners. Guests bowled and played shuffleboard and tennis. Carnegie put friends up and hosted parties there. He read books and went on nature hikes and bird-watching walks. He engaged in elevating conversation. On the mountain above Johnstown, Andrew Carnegie, presiding spirit of his entrepreneurial age, leader and model of the Pittsburgh industrialists who were turning the whole region into a scarred, smoky mess, was taking up the growing fad for simple outdoor pleasures, clean air, and contemplation of nature. It was in following Carnegie's lead, and then inducing him to join their organization, that in 1879 the members of the South Fork Club determined to dam up a river and make a beautiful mountain lake.

CHAPTER 2

DOWN IN THE VALLEY

Daniel J. Morrell wasn't only a Johnstowner. He was Johnstown itself.

And Daniel J. Morrell's approach to business was directly at odds with that of Andrew Carnegie. So too was his attitude toward pleasure.

When in 1879 the men of the South Fork Club first began creating a lake up on the mountain above his city, Morrell, two hundred pounds of stocky, old-school Quaker businessman, factory manager, and civic leader, took a decidedly dim view. By then Morrell had come to know those Pittsburgh men. He saw how they operated. On the Andrew Carnegie model, they didn't like to get too closely involved with what they were doing. Their wild success had come not from doing things themselves, but from putting money into, cutting costs in, and taking over things other people were doing. Morrell understood that these rising industrialists were financially ingenious and economically and politically powerful beyond anything seen before. And partly for that very reason, he mistrusted their ability to build a good dam.

He doubted their ability to do that also because he'd already tussled with none other than the upstart Andrew Carnegie himself, and as powerful as he was in Johnstown, and throughout the region, Daniel J. Morrell had lost that battle. But he wasn't down-and-out; he had all the clout to be had in Johnstown, and he didn't like the Pittsburghers' dam project on the mountain above his and his people's heads.

Born in North Berwick, Maine, in 1821, Daniel Morrell moved to Philadelphia in his teens and took up work as a counting-house clerk. By the mid-1850s he'd become connected to a group of Philadelphia investors in the Cambria Iron Company, a venture first formed by an eastern consortium in the 1840s. The Cambria venture's purpose had been to operate an iron furnace, intended to be the biggest and best among the many smaller furnaces that had begun to fill in the western Pennsylvania countryside with the decline of ironmaking in the Juniata River Valley to the east. As early as the 1840s, when not much big industry was going on there, the original Cambria consortium had its eye on Johnstown, Pennsylvania.

Why, of all places, Johnstown? The usual three reasons: location, location, location. John Roebling, later to engineer New York City's Brooklyn Bridge, was working as an assistant engineer for the state of Pennsylvania in the 1840s when he reported that in Johnstown, Pennsylvania, "capitalists could hardly find a more eligible situation for starting mammoth furnaces on the largest scale." That might have seemed a funny description of a place considered remote until recently. The first white settlements of western Pennsylvania had come somewhat late, compared even to settlements in the western regions of Massachusetts or Virginia. In 1776—only seventy years before the first Cambria people started getting into iron—Pittsburgh itself was a mud-street village straggling around the collapsing, often flooded Fort Pitt, from which the British Army had commanded the strategic point at the Ohio headwaters. The colonies of Pennsylvania and Virginia had been fighting over Fort Pitt—it was literally at times a shooting war—and over which of the two colonies that village lay within. The fight was strategic. At Pittsburgh the Allegheny and Monongahela rivers flowed together to form the headwaters of the Ohio River, opening the way to what many believed, even in the 1770s, would someday be a bountiful American West.

But even that muddy village of Pittsburgh had seemed, back then, somewhat civilized compared to the forbidding mountains just to Pittsburgh's east, up in that great, barely passable Allegheny barrier separating the remote Ohio headwaters from the rolling farmland and the eastern seaboard. The first white settler in the wild upland area then called Laurel Hill—a general name for what was to become the mountains above the Johnstown region—was a man named Herman Husband, and when it came to the kind of hard labor and concentrations of wealth that would dominate the region in the nineteenth century, Husband set an early tone of protest. He wasn't a laborer but both a rich planter and a fugitive from British-colonial justice. He'd led an uprising in North Carolina, ten years before the American Revolution, on behalf of ordinary small farmers, artisans, and landless laborers, against entrenched elite interests sewing up trade in the North Carolina backcountry. Those rebels called themselves "regulators," because they wanted to regulate big business and big money. Herman Husband led them in a military action against the Crown and the North Carolina colonial government.

Leading that uprising was a capital crime, and when his rebellion was put down by the governor's troops, Herman Husband adopted a memorable alias—"Tuscape Death"—and fled northward into a place he knew he'd never be found, that steep, barely settled upland just east of the Ohio headwaters.

He was right: nobody found him there. And once the revolution validated his freedom, Herman Husband again took up the cause of his poor neighbors, the incoming settlers of western Pennsylvania. But now instead of the Crown he was fighting the American elites in Philadelphia. Representing the western highlands in his new state's assembly, he used what little electoral clout he had to push for greater democracy and economic equality for the small farmer, the artisan, and the laborer.

But America's first western land boom was on—the Alleghenies and the headwaters were the original American West—and big investors from the East were quickly buying up huge tracts of the region that nobody had wanted when Husband had first moved there. Local rich people, too, were on the make: connected by cronyism to Federalist politicians back East, they had government and army appointments in the western outposts, and soon it seemed to many ordinary people in western Pennsylvania that a few closely interconnected local families were using their privilege to "engross," as the term then had it, all of the best land. The rich foreclosed loans on struggling debtors and turned once-independent settlers into their own day-laboring farmhands. Small-scale factory work was already beginning then, too: commercial distilling, glassmaking, pottery, shipbuilding, gristmilling, and other industries used waterpower from the many fast rivers. The rich families sewed those businesses up, too. By the 1790s, flatboats and other craft were plying the Ohio, and wealth was booming for the federally connected absentee land speculators and local waterpower industrialists alike. Ordinary people of the region had less and less ownership. They were already becoming workers on behalf of others' wealth.

So it was that the first major American action on behalf of small farmers, independent artisans, and laborers, in opposition to the well-connected elites' control of wealth, occurred in 1794 in those same mountains and valleys east of Pittsburgh that would one day become the center of operations for the industrial giants of the Gilded Age. This was in 1794: Herman Husband was in his seventies now, but once a regulator, always a regulator. Naturally he became one of this rebellion's leaders.

In what was known as the Whiskey Rebellion, because it reacted violently to a federal tax on distilled spirits that ordinary people in western Pennsylvania—the best whiskey was made there—viewed as directly penalizing them on behalf of rich people, both locally and nationally, the rebels demanded what they called equal taxation instead. They massed in the thousands on the banks of the Monongahela, in the place called Braddock's Field, where the French and Shawnee had defeated British general Braddock years before. They flew their own flags of secession. They had a shoot-out with federal officials. President Washington personally led twelve thousand troops across the Alleghenies to establish control, and when the president turned back for Philadelphia he left the operation to his Treasury secretary Alexander Hamilton, author of the tax law they were resisting. Subjected to Hamilton's strongarm enforcement methods, the people of western Pennsylvania soon got the message. Federal authority was asserted in what was then the American West.

When Daniel Morrell's Philadelphia partners decided to get into iron in Johnstown, only sixty years had passed between the time President Washington sent an army to the area to establish the United States' sovereignty there and the beginnings of modern industry in the region. In that brief interim, the Johnstown region had changed, to put it mildly. Modernity had occurred, seemingly all at once, and Johnstown had become a canal town. John Roebling was right to view the flat site below the mountains, where the Little Conemaugh and the Stony Creek flow together, as an ideal site for capitalists in the iron business to set up shop.

In 1852 the Pennsylvania Railroad established through service from Philadelphia to Pittsburgh, with a convenient station in Johnstown, making the canal, so important but so short-lived, obsolete. Now that raw materials could be brought easily, product moved out just as easily. Johnstown's position at the point where the Conemaugh becomes navigable meant product could also handily be loaded onto flatboats and go right down into the Allegheny River to Pittsburgh, onto the Ohio River, all the way to the Mississippi and New Orleans.

The Cambria Iron Company, first founded in Johnstown in 1852, had nevertheless been struggling under its first owners. So the big, gruff Daniel J. Morrell and his Philadelphia partners began by leasing its works for their own enterprise, and that's when Morrell, appointed general manager, first came across the Alleghenies, on a mission to find out what was wrong with the company and fix it.

On arrival in 1855, fix it he did. Morrell introduced efficiency and reliability and started making a profit. With that going, Morrell and his partners identified a major opportunity. They started buying up Cambria company bonds. When the company defaulted, in lieu of payment they took over the whole thing, no longer leasing but owning it.

Daniel Morrell and the Cambria Iron Company transformed Johnstown. The city boomed on the Morrell partners' Cambria iron. Even as the Pennsylvania Railroad was putting the canal system out of business, Morrell not only turned Cambria into the monster set of mills for which it would become famous, but also remade the city as one of the first of the classic company towns. With Morrell as their leader, a small group of Cambria Iron Company men—over time they would intermarry into a tight clan—built Johnstown up and controlled it, too.

Morrell wasn't just the head of the Cambria company: he was president of the Johnstown Gas & Water Company, president of the First National Bank of Johnstown, president of the Johnstown City Council. He and his Cambria people thus ran banking, the streetcar, the streetlights, everything that most of the other towns in the region lacked, and which gave the people of the city—the middle class anyway—such pride and enjoyment. By then Daniel Morrell lived in the biggest and best house on Johnstown's Main Street, brick, with a mansard roof. His lawns and gardens, encircled by a high fence—made, naturally, of iron—took up a full block.

He brought in skilled workers: the plumbers, the pipefitters, the Welsh miners, the masons, the carpenters. He brought in unskilled workers, or those immigrants whose foreignness relegated them, regardless of skill, to ranks of the less well paid. By the end of the 1860s, Cambria had four thousand employees, causing the boom not only in iron, and then in steel, but also in mercantile enterprise that was making James Quinn's family so prosperous.

"Company town" became a literal term, even beyond city government, finance, and essential services, when Cambria built eight hundred small frame houses for its lowest-paid workers, ran a company store, and built a company school, a company hospital, and a company library. Cambria Iron was soon the biggest landlord in Johnstown. And it built the city's opera house.

All of this abrupt, ongoing development involved Morrell's ceaseless, hands-on innovation. Cambria Iron was the first works in America to use what was known as the "retort" coke oven, which made coal into coke with high efficiency, and Morrell was always seeking new ways of boosting the power of the locally mined, low-volatile coal, so conveniently nearby. He opened the company's own local coal mine, and thus avoided the costs of rail transportation.

Then Morrell went further. He beat Andrew Carnegie into the steel business. And it was in his competitive relationship with Andrew Carnegie, and the other emerging steel men of Pittsburgh, that Daniel Morrell developed a sense of why the lake above Johnstown might not be ideally held back by the dam the South Fork Club had built.

As much as iron had changed Johnstown and the entire valley, steel now accelerated that change. Because the region would remain associated with steel for more than a century, and because the first name in American steel was soon to become that of Andrew Carnegie, many would assume that the industry was pioneered in Pittsburgh. But the steel industry began not in Pittsburgh but in Johnstown, and Daniel J. Morrell and the Cambria Works were its most dynamic pioneers.

It was an ingenious new thing, industrial steel production.

Steel, like iron, is an ancient product, but while mass-produced iron had led to great fortunes in a new industrial age, producing steel on a similarly industrial scale had long seemed impossible. Steel had existed for thousands of years, and for most of that time its makers knew only that under certain conditions, and using certain hands-on, artisanal processes, a metal at once far harder and far finer than iron could result from variations on the usual iron smelting process. They didn't have the chemistry to know that steel actually results from reducing the carbon content of iron.

So they groped at steelmaking. In ancient chimney furnaces, iron was refired: that removed carbon and achieved steel. Combining iron with less carbon-heavy elements also served to make steel. The result was prized especially for its capacity to be sharpened to the razor fineness desired in weaponry, but well into the nineteenth century the process of turning iron into steel remained small-scale and labor-intensive.

Steel swords were one thing. Steel rails and bridges and buildings were hard to imagine. For a good part of the industrial revolution, mass-produced, workable iron seemed the dominant metal of modernity, the driver of fortunes, the purpose of factory labor.

Then in 1855, when young Andrew Carnegie was still working in mills, Henry Bessemer, an English inventor and munitions maker, eager to reduce the price of steel production, found a quick, reliable, large-scale process for removing carbon from pig iron, using a readily available resource: air. The Bessemer process was revolutionary—and surprisingly simple. Blowing air through iron in its molten state puts oxygen into the mix, and the resulting oxidation of the hot, liquid iron isolates carbon, releasing it—yet another kind of slag—to the bottom of the mix, leaving above a soup of nearly pure steel.

Earlier steelmakers had figured out this air trick. Bessemer-like processes had gone on in China and Japan from the eleventh century. But Bessemer was a nineteenth-century industrialist. After a number of false starts he was able to get the whole process controlled for procedure, quality, and especially for the scale enabled by modern factory methods.

The Bessemer process occurred in a big iron tank, and "the blow," as it was called, took only about twenty minutes to produce a pure, molten steel ready to be ladled into molds. There it could be mixed with other elements to further harden it, and make various kinds and grades of good steel. Where earlier methods had yielded about fifty pounds of steel in about fourteen hours, Bessemer could make tens of thousands of pounds in about half an hour.

That changed everything. Performed on a mass scale in a new kind of factory, the steel mill, the Bessemer process reduced the cost of steel production by about 80 percent, putting the superior metal, shockingly quickly, at about the same cost as wrought iron. The first U.S. steel mill was built in Troy, New York, in 1865, on a license from Bessemer. In 1867, the process was exhibited to the American public in Chicago. Now everything moved fast—nowhere more so than Johnstown, Pennsylvania.

As chief executive of the Cambria company, Daniel Morrell was free and willing to spend a lot of money. He'd already financed a series of costly pre-Bessemer experiments in industrial steelmaking. They weren't successful, but the very fact of Morrell's willingness to spend on new ideas brought the best steel minds in the country to Johnstown. The Fritz brothers, George and John, built Morrell a Bessemer mill in 1867. A three-roller process for cutting out steel rails was pioneered by Bill Jones, sometimes known as Captain Bill, one of the great steel men of the age and, as important, also a hands-on, well-liked leader of skilled steelworkers, a man who led from the front, not the rear.

It was with the help of Captain Bill Jones that Daniel Morrell made the critical, nontechnological innovation that advanced the Cambria company to the head of the steel pack. When the first steel rails in America were rolled experimentally in Chicago, ironmasters believed that the Bessemer process could be carried out only by skilled workers brought over from England.

But Daniel Morrell, the Fritz brothers, and Bill Jones thought differently. They hired American workers, Bill Jones trained them, and in 1867 it was the Cambria Iron Company of Johnstown, Pennsylvania, that delivered the first set of Bessemer steel rails commercially ordered in America. They were made by American labor.

That was still small-scale, and Morrell kept spending and experimenting. In 1871, Cambria built a huge, state-of-the-art Bessemer plant and took its place as the country's biggest producer not only of iron but also of steel. By the end of the decade, seven thousand men were employed in four furnaces in one plant, two more furnaces in another factory, the huge Bessemer plant with its three-ton converters, and soon an "open-hearth plant," using an even more advanced steel technology. There was a rolling mill two thousand feet long. There was a machine shop for making replacement parts for the machines. It all went on in long shifts, night and day.

Also under Morrell's management, Cambria verticalized. Cost-cutting wasn't his trick: he controlled cost by controlling the supply chain. Cambria bought up thousands of acres of real estate around town where coal was abundant, further enhancing its ability to mine its own raw materials. The company owned a local works to turn that coal into coke, thus providing its own fuel supply. To add value to Cambria steel at the other end, Morrell invited the Gautier Wire Company into town. Thousands of miles of new fence was being strung up throughout the county: barbed wire, made possible by steel. Morrell personally owned half the Gautier company's stock, and soon the Cambria company had taken it over completely. Running on coal sent straight down a chute from the hills, Gautier was making thousands of tons of barbed wire per year out of Cambria steel, along with rods, tools, and other steel products.

"Cambria-land," some would later call it. Approaching by rail from Pittsburgh, a visitor's view east was blocked by a curtain of smoke from the mills. The sheer, raw ugliness was breathtaking. A natural gas pipeline ran from Greensburg to feed the mills. Stacks belched smoke and fires gleamed. Four red-brick blast furnaces went day and night, throwing up geysers of sparks, while the incredibly long, monotonously faceless rolling mills boomed and the coal trains rumbled below partially stripped hills.

The ugliness meant that, by 1876, the company was debt-free. It had an inventory of 75 million tons of fire clay, 20 million tons of iron ore, and 350 million tons of coal.

All of that was accomplished by American labor. But not by union labor. Daniel Morrell's way of identifying the city of Johnstown with the Cambria company meant obstructing anything that challenged the company. He saw labor organizing as one of the major threats.

Organizing would seem a natural outgrowth of the labor itself. Inside the Johnstown steel mills, images of hellishness were fully borne out. It was hot, to put it mildly, on the floor, which would have seemed chaotic to a visitor. But amid the sparks, soot, and noise, a complicated system was constantly in play. Coal was burning orange in the darkness of a booming, resounding, crowded, cavernous hall. Men were shoveling coal into burners for the blast furnaces that were turning iron ore into molten pig iron and smoldering piles of slag; shoveling coal into burners of oxidation tanks, turning the pig iron into molten steel and yet more slag; shoveling coal onto the newer, open-hearth furnaces that were starting to complement the Bessemer process. Men drove small engines fast on curving rails, powered by steam driven by more burning coal, took slag away, and dumped it. They removed steel from the hearths in long-handled ladles and carried the liquid metal to various destinations, including steam-driven rolling mills, which, when long bars of the still-soft steel were pushed through the rollers, carved the bars into rails for train tracks.

Shifts were twelve hours long, in some factories seven days a week, often with no time to eat or even take a break, except for those few minutes when a machine had to be oiled. The men poured sweat, in part because despite the brutal heat, they sought some measure of physical protection from injury by wearing two layers of long wool underwear. They drank two buckets of water a day and stayed parched. Sweat slicked their hands and filled their shoes. Their faces blackened by soot, the men were burning calories and losing water so fast they were nearly as skinny as the rails themselves.

Nimbleness was key to survival at all times, but when exhaustion late in a shift took hold, injury and death occurred with alarming frequency. Slag piles fell and buried the men, ladles tipped molten steel on them, iron and steel beams cracked their skulls, coal seared them, steam scalded them. A factory whistle, blowing at a random time of day, meant that such an accident had occurred, and families in town froze in horrid anticipation of who it might have been and how bad it might be. With some injuries, death would be mercy.

If you survived such six-day weeks until the age of about forty, having kept your family barely above the poverty line at about ten dollars a week, you were old, finished, with no view of further income. You'd be dependent on the next generation.

Fortunately or unfortunately you probably had too many kids for comfort. Family life was crowded. Housing for the laboring families living in Johnstown and the nearby coal towns around the Conemaugh River often consisted of purpose-built

structures, hastily thrown up by management to accommodate as many people as possible as cheaply as possible. Identical, small, two-story houses in tight rows, or sometimes fully
attached, lined rows of long, stark, treeless streets, often strewn
with rubble. Getting the essential fuel, coal, out of the ground
drove everything: the mines were dug as close to the steel mills
as possible, and the coal towns, in turn, abutted the mines and
the rails; trains rattled and groaned sometimes only a few feet
from the houses. Shared yards behind houses sometimes offered
the many children the only place to play. Laundry flapped on
lines there, darkening in the soot.

The number of children and the endless laundry gave testimony to the lives of women whose husbands, brothers, and sons
worked in the mills: a never-ending, seven-day cycle of up to
eighteen-hour days of domestic work, on desperately tight economic margins, as exhausting as any factory job. Women and
girls also ran their cramped homes as boardinghouses for further
income and planted and harvested vegetable gardens.

This grueling human labor was extracted at debilitating levels, all in the service of efficient steel production. And it was
compensated at a barely subsistence level, for maximum profit.
Despite the ingenuity of the new technologies, the brilliance of
the investing, and the resulting national dynamism, human
labor remained the primary natural resource fueling the entire
enterprise.

The upside for the worker was simply employment, which
beat starvation. Small agriculture and artisanal production were
becoming things of the past in western Pennsylvania. In 1850,
farms were still where most Americans worked; in 1900, the
majority of the nation would work in factories. Wage labor—in
its most efficient form, factory work—was what gainful employment was largely coming to mean for great numbers of ordinary

American families by the 1880s. This nearly impossible way of life was pretty much all there was for a huge number of people.

Since conditions in Europe were in many ways far worse, more and more people, seeking this American way of life, were arriving in western Pennsylvania, just as the Carnegies had in the late 1840s, drawn by the industrialization that led to the boom in factory employment. Residents of the Conemaugh Valley were by no means solely the Americans largely of English, Scottish, Scots-Irish, and German extraction who had risen up against the federal government and its local cronies back in the 1790s. The German American population had come to dominate Johnstown by the 1830s—the town had been founded by a German named Schantz—but by the 1870s the mills and mines were drawing Welsh coal miners; Irish laborers originally hired as railroad workers, now hired by the steel mills; new Germans straight from Germany; and Scandinavians.

Those were western European immigrants. Bosses and earlier western European immigrants considered them more or less civilized, and the better-paid factory jobs went to them.

Jews too began arriving in the Johnstown area around 1850, fleeing pogrom: first German Reform, then the Orthodox from eastern Europe. But they were barred from most jobs and not to be hired by the mills. Unlike many of their more urban counterparts, Jewish Johnstowners therefore operated in the local economy in a close-knit entrepreneurial niche, maintaining religious congregations while serving the larger community, mainly via shopkeeping.

Then came the Gentile eastern and southern Europeans: Italians, Poles, Russians, Bohemians, Hungarians, Slovaks. Unlike the Jews, they were hirable for work but were relegated to the least-skilled, lowest-paid jobs. A lot of these new immigrants were men alone, having left families behind in hopes of returning home with cash or bringing their families over after settling in. Most ended up staying.

"Hunkies," they were called—a slur for Hungarian, but applied indiscriminately to all eastern Europeans—or "Bohunk," a portmanteau slur including Bohemia. They were shunted out of Johnstown to live in Cambria City and Minersville, places less convenient to the work. There the conditions were even worse, and as such towns became up to 85 percent foreign-born, they were known, and scorned, by the Johnstown people as Hunkie-towns.

In some ways, the Cambria company's furnaces and mills seemed ripe for organizing this mixed pool of labor to bargain as a union for better conditions, maybe even for higher pay. Unions did exist. Organized labor had begun in earnest as a response to industrialization. In the nineteenth century, when greater numbers found themselves dependent on wages, booming industries from rail to coal to steel became increasingly dependent on great numbers of ceaselessly working people, and improvements in automation enabled industrialists to hire more and more unskilled workers, both workers and activists began to see that the only control they might ever be able to exercise over their destinies and liberties would be to organize. No one person or even small group could hope to bargain for improved conditions in a mine or steel mill. No laws protected employment, and unlike in a feudal system, or in the apprentice system—notoriously unequal, of course—factory owners had literally zero reciprocal responsibility for the lives of workers. With nothing but one's own labor power to sell or withhold, only collective bargaining—agreement by a large group of employees to provide a great mass of labor power in exchange for concessions by owners—could ever create improvement for workers.

So throughout American industry, conflict between labor and management had been heating up. The largest outbreak of the 1870s, presaging bigger things to come, occurred in 1877 in what became known as the Great Railroad Strike. It shook up the bosses: this wasn't just a strike but an uprising, and it was widespread, like the railroad, disrupting service and causing deep financial losses in many places and for many rail companies in overlapping waves of protest. In Pittsburgh, the workers on the Pennsylvania Railroad reacted to the company's announcement of both a 10 percent pay cut and an increase in the number of "double-header" trains: the use of two locomotives, each with its own crew, at the head of a train, a risky practice demanding highly skilled crews. The workers seized the Pittsburgh depot and blocked rail traffic in and out. Freight trains were stuck, and the strikers gained support from people from all over the city, running down to the train yards to join in the action. Every working family in Pittsburgh, train workers or not, hated the Pennsylvania Railroad.

That strike and uprising ended the way so many more would, as well: in a military suppression by the armed forces of the state. Because many in the Pittsburgh militia seemed to be in sympathy with the strikers, troops were called in from Philadelphia. The crowd of men, women, and children was armed with rocks, bricks, and some pistols, and in a series of chaotic actions, troops making a bayonet charge on the crowd started shooting. The battle for Pittsburgh raged on for days, and left 61 dead, 124 injured, and 139 arrested, but in the end the company and the state were victorious.

The Cambria Works paid its workers less than almost any other Pennsylvania steel company. Yet the Amalgamated Association of Iron and Steel Workers, the union for the industry's skilled labor, couldn't get a toehold there. Morrell had the whole town and the Cambria Works at once under his thumb and in his care. Organizers getting off the train in Johnstown were met by police suggesting they take the next train home or go to jail. At the same time, the company really did provide many of the civil services, at a high level of quality, on which ordinary working people depended, and did so far better than any other town in the region.

That was Morrell's way. He took care of his people, and so the people were his. In 1874, before the advent of industrial steel, a powerful skilled-ironworker group called the Sons of Vulcan did try to strike Cambria Iron. They got nowhere. That same year, the company crushed a strike in its coal mine, and Cambria issued a set of rules outlawing union participation by its workers. Anyone joining a union would be fired. It was that simple.

The Cambria company flourished, and yet by 1879, when the South Fork Club started building its dam, Daniel Morrell had experienced setbacks. Not only financial losses—the steel business went wildly up and down in the mid-1870s, following a financial panic in 1873—but also loss of control. That diminishment had come at the hands of Pittsburgh's biggest man, Andrew Carnegie.

Carnegie was never, like Morrell, a hands-on experimenter or manager. At his Union Iron Mills and Lucy Furnaces, Carnegie's brother Tom had run the business, and Andrew Kloman had supervised the work and the workers. So when Andrew Carnegie realized the future was in steel, and that he had to own that future, he knew he had to make some characteristically sharp moves.

Erecting his own sprawling, red-brick steel works in 1872 in Braddock, Pennsylvania, just up the south bank of the Monongahela from Pittsburgh—ironically enough, this was just where members of the Whiskey Rebellion had mustered in 1794 in opposition to concentrated wealth—Carnegie named it the Edgar Thomson Steel Company, after his old railroad boss and inside-deal partner, and set about competing with the leading Cambria Works, run by Daniel Morrell. This mode of competition began with raiding Cambria's best minds and most skilled workers.

Carnegie always knew whom to hire, and so at first of course he tried to entice Daniel Morrell himself to leave the Cambria company and leave Johnstown, and come to Braddock and run the Thomson Works. He offered Morrell $20,000 per year and some shares.

Morrell turned him down. The gigantic, old-school Quaker viewed the slim newcomer Carnegie as a flighty speculator, a social butterfly, and an absentee owner, no committed ironmaster and steel man but a mere financial game player. He thought Carnegie lacked principle, in that the upstart seemed to be making money not from work but from money itself. Morrell had little confidence in the future of the Thomson Works. That was an error, to say the least.

Spurned, Carnegie bided his time. Now he had his eye on Bill Jones, Captain Bill the steel foreman, who had been with Morrell for sixteen years now and had first made the Bessemer process work at Cambria. Knowing nothing about making steel, Carnegie needed the best Bessemer man in the business, and he knew that man was Bill Jones.

Luckily for Carnegie, some tensions had meanwhile arisen among the management of the Cambria Works in Johnstown. Having hired and trained the American workers Morrell had insisted on hiring instead of British specialists, Bill Jones had good reason to expect a promotion. Yet when the Cambria manager died, Morrell passed Jones over and hired somebody else.

Another error. Carnegie pounced.

In 1875, Carnegie offered Bill Jones a job managing the Edgar Thomson Works. The decision was a no-brainer for Captain Bill. Along with Bill Jones came his personally trained crew of skilled steelworkers, the men who actually made the process happen. At Carnegie's Thomson mill in Braddock, Jones and his men went right on innovating. Over the years, they designed power-driven tables to move the rolling machines, drastically reducing the number of men needed to handle the rails. They increased the number of times a rail could be passed through the rolls before it needed reheating. Production went up, along with Carnegie's favorite thing: cost savings.

Within a year of Jones's move from Cambria, the Edgar Thomson Works was beating Cambria at production. Carnegie was high on a mountain—not in the Alleghenies but in Italy— when he heard that news. On receiving it, the industrialist actually did a little dance of joy.

And thanks to Jones, things just kept getting better for Carnegie's plunge into steel. As the 1870s came to an end, Carnegie was consistently beating Morrell on output. Suddenly Cambria was not the country's biggest mill. Carnegie's was. Even as Carnegie laid plans, ultimately successful, to dominate a big piece of the American steel industry, it was Captain Bill Jones who, rightly enough, began to gain world renown as a steel man. Jones was famous in Pittsburgh and far beyond. He gave talks and published papers.

Yet when Carnegie offered him a partnership in the steel empire, Jones turned him down. If he became management, Bill knew, his relationship with his men would be over. Jones demanded a high salary instead, and Carnegie offered him what the president of the United States made: $25,000 per year.

◆ ◆ ◆

Meanwhile, Carnegie was using his connections and insider adroitness to further undermine the Cambria company's dominance in steel. Leveraging his relationship with his old mentor and friend Tom Scott of the Pennsylvania Railroad, Carnegie got Scott to pressure Cambria to take a step back and let Carnegie into the steel rail business.

The railroad bought a lot of rail. Competitive bidding wasn't part of the process of determining whom the company bought its steel from. Tom Scott now told Morrell that Cambria must form a cartel with Carnegie and split up the railroad's steel business. Cambria and the other major competitor, Pennsylvania Steel, couldn't supply all the railroad's needs anyway; still, Scott implied that if Cambria and other competitors didn't allow Carnegie a chunk of the railroad business, they'd lose it all.

The steel companies met in Pittsburgh, under Morrell's aegis, and divided up the business. They had no choice. Morrell's ideas about how Carnegie operated were being borne out.

In other areas, Carnegie went directly after Cambria, using sharp tactics that infuriated Morrell. During a time when steel prices underwent a general and precipitous drop, there was a certain price below which Morrell would not go. Carnegie, however, was willing to sell at a loss: he figured the drop would be temporary, and that the longer Cambria didn't fill orders, the bigger the chunk of market Carnegie could capture. He was in the investment business, not the steel price business. Carnegie told Morrell that he would agree to fix prices in collaboration with Cambria, and that unless Cambria agreed to fix prices, he would fill orders and stay in business, regardless of price. Morrell knew iron and steel. Carnegie knew capital, and Carnegie was taking over. So Morrell did agree to pooling and price-fixing. Cambria and Carnegie were now in a kind of monopolistic partnership that went against all of Morrell's competitive instincts, especially since Morrell had become the weaker player. He couldn't see the future, but he realized he couldn't compete with the Edgar Thomson Works of Pittsburgh.

Still, just because Morrell couldn't go head-to-head against the Pittsburgh capitalists in business didn't mean that he, or Johnstown, had to put up with what they did for fun in the Alleghenies. Daniel Morrell had plenty of power left, and he was full of disdain for the Pittsburghers' abilities when it came to anything other than making money.

Fresh air and exercise. This was the new idea. Carnegie had taken it up, and by 1879, when a group of fabulously and somewhat less fabulously rich Pittsburgh industrialists decided to form the South Fork Club, it was becoming something of an American fad. The fact was, the world had begun to smell bad, and people with money had the ability to get their families out of that unpleasantness. Nobody who didn't have to put up with it, especially in the hottest days of summer, was going to stick around.

To cool off, breathe fresh air, and get back in touch with nature, people who could afford it went to places like Bar Harbor, on the coast of Maine; Newport, Rhode Island; the lakes of the Adirondack Mountains; or the Tuxedo Club, north of New York City (founded by tobacco magnate Pierre Lorillard, Tuxedo lent its name to the formalwear preferred by its well-heeled members). In Pittsburgh, where many of the members of the South Fork Fishing and Hunting Club had their palatial homes, the befouling of the air by ceaseless industry was perhaps the worst, and for some time now the new-rich magnates of that city had looked with envy on summer refuges enjoyed by the older-money rich of the East.

The Pittsburghers were landlocked. Nor did their mountains have the awesome scale or high, rolling regions of a romantic wilderness like the Adirondacks. The magnates did have the money to travel to Lake Placid and Bar Harbor, and they did so, in splendor, in their private railroad cars, with their families and servants and pets, and even with their china and furniture stowed in baggage with their trunks and trunks of clothes. But they wanted something like that of their own.

Andrew Carnegie showed them the way. Living, when he was in Pennsylvania, not in Pittsburgh but up at Cresson, he endowed the Alleghenies with prestige. In the mountain passes above the roaring industrial towns of western Pennsylvania, so convenient by rail to Pittsburgh, the land was still wild and precipitous and beautiful, and the air was clean. All you had to do to get away from the industrial pits was be conveyed, first by rail, and then by carriage, upward. In the forests, birds still sang; in the fields, cows still grazed. Just looking up from Johnstown, in certain directions you could still see it all, pasture and forest, the way things used to be, right above the city.

And so in 1879, some rich families of Pittsburgh formed the South Fork Fishing and Hunting Club. It was to be their own Cresson, yet even better than Carnegie's retreat. For one thing, the Mountain House, where Carnegie was content to put up guests and hold parties, was just a commercial railroad hotel. Anybody could stay there; it had no privacy. Even Carnegie's house, so near the grand hotel, seemed a bit open to gawking view: Cresson lay right above the railroad, giving easy access to ordinary commercial tourists and holidaymakers.

And the pleasures to be enjoyed in the fine air at Cresson were limited. These magnates wanted the spiritual and physical restoration that Carnegie and other faddists were always going on about, but to sportsmen like them, with their growing families in need of social lives and athletic outlet, restoration meant not just bird-watching and reading and high-minded talk, but hunting and fishing and boating. In forming the South Fork Fishing and Hunting Club, the big men of Pittsburgh came up with a restorative escape more ambitious than Carnegie's home in his mountain aerie, more private than the Mountain House, more amenable to the popular upscale sporting activities of the day. Regardless of what Daniel Morrell might prefer, they would have a lake, and Carnegie would join them there.

CHAPTER 3

HOW TO MAKE A LAKE

While Carnegie, making his untold millions in rail, oil, telegraph, iron, and steel, lived part of his mental life in poetry, literature, and rejuvenating contemplation of natural beauty, it was Benjamin Franklin Ruff, far less famous than some of the other members of the South Fork Club, who actually conceived of creating a mountain retreat at the crest of the Alleghenies above Johnstown, bought the land, started improving it, and sold shares in the enterprise to his mighty friends among the captains of industry. And it was Benjamin Ruff who dammed up the South Fork Creek.

Benjamin Ruff was no Andrew Carnegie, but that didn't make him unusual among the big men of Pittsburgh: nobody but Carnegie was Carnegie, though everybody had to do business with him, and Ruff was no exception. An important figure in booming industrial Pennsylvania, intense-looking and peremptory to an extreme degree, with a black beard, he was the same age as Carnegie, having begun life in 1835, in Schenectady, New York, and like many others he'd risen to prominence in western Pennsylvania doing business with the railroad, in Ruff's case as a tunnel contractor. He'd also made money selling coke and brokering real estate deals.

Now he envisioned a gorgeous summer retreat up on the mountains. Here the great men of the day, along with those who had enough money to style themselves great, would escape the cares of business and find peace and quiet at play, far from the prying eyes of the hoi polloi who worked for them below.

In addition to increased privacy, the new club's most important improvement over Carnegie's Cresson was to involve fishing and boating. Cresson had springs, of course; its full name was Cresson Springs. But it had no lake. How could it? Cresson was nearly at the top of the mountains, and the Allegheny crests don't offer those rare craters where, in some other regions, mountaintop lakes may sometimes exist. The Alleghenies form a big part of the Appalachian divide: the runoff drains, via many rivers, east to the Atlantic Ocean, west to the Ohio, and then to the Mississippi and the Gulf of Mexico. That water doesn't stay around. There are no lakes.

But the South Fork Fishing and Hunting Club would have a lake, and a beautiful one at that. If nature couldn't supply the Allegheny crest with such an important feature, then Benjamin Ruff would simply have to do it himself.

The Pittsburgh millionaires' deep desire for a lake responded to certain recreational crazes just then taking off among the well-heeled, not only in Pittsburgh and environs but around the country. These sport crazes in one way aligned nicely with Carnegie's ideas about the spiritually regenerative powers of recreation, natural beauty, and contemplation. Yet they also went well beyond such simple prescriptions. These sports called for a lot of gear.

Black-bass fishing, for one. While fishing had long been the consummate gentleman's hobby—Izaak Walton's classic *The Compleat Angler* was published in 1653—those traditions referred mainly to fly fishing for trout and salmon, the sport of Scottish lairds and those who wanted to feel like them. There was a certain elegance to fly fishing, and wealthy Americans had taken it up at least as early as the middle of the century. Fly rods were long and took expert handling in order to get the light "fly"—it might be one of many artistically tied creations masking a hook—looping back and forth through the air before landing with a gossamer touch on the water's surface, to bring a trout eagerly seeking a bite onto the hook. That was dry-fly casting; wet-fly fishing, less impressive, involved submerging the fly and waiting for a bite. Either way, you might make a little campfire and roast one or more of these trout, or bring them home to fry for breakfast, but eating wasn't the point of the exercise. That's what made angling for trout a gentleman's sport.

Black-bass fishing, by contrast, had generally been carried out for purposes of getting food. That meant it was done by poorer people, using not extra-long, whippet-thin rods or fancy, hand-tied artificial bait but short, sturdy poles and live bait, usually worms and minnows. And yet by the middle of the nineteenth century, American sportsmen, in a kind of Yank rebellion against the complicated pretensions of British fly fishermen, developed a new sport out of the hunt for the black bass. As the Gilded Age began, bass fishing competed with and sometimes complemented trout fishing in the pantheon of the manly pastimes of the upscale.

So bass fishing too now had its artificial lures, not flies but glittery tin or brass spoons for trolling from the stern of a slowly rowed boat, fake minnows for casting. And to the bass-fishing gearhead of the day, these lures took on some of the mystical properties that fly fishermen often ascribed to flies. Long debates about tackle and gear could while away many an hour over a post-angling drink or three. Nor were upscale bass fishermen allergic to the live-bait approach taken by the working-class originators of the game. Sportsmen too now baited their hooks with real min-nows, grasshoppers, frogs, and saltwater shrimp tinned in brine.

Not worms, though. That would be taking things just a step too far.

Along with bass fishing, there was another water sport re-quiring a lake, when Benjamin Ruff was seeking out land for his hunting and fishing club: sailing. The New York Yacht Club, the first American organization devoted to sailing not as a means of conveyance but as a sport, had been founded in 1844, and while racing big boats on the ocean was an expeditionary adventure, small-boat sailing had come into vogue on the wide, windy lakes of the Adirondacks and the White Mountains where the wealthy decamped for the summer. Young people especially could be seen heeling their boats at extreme angles, running the big lakes in a thrill of sheer speed.

That wasn't something you could do anywhere near Pitts-burgh. Lakes were scarce not only in the mountains but every-where in the state, and rivers throughout the region had long been dirtied, dammed, and beaten up by oil and mining proj-ects. You just didn't find clean lakes filled with fishermen and fast-flying sailors enjoying pure mountain air. But Pittsburgh's elite wanted to relax and recreate as easterners did, and Benja-min Ruff was going to see that they could.

He knew just where the right land lay. Farther than Cresson from the railroad, with its hoi polloi, yet still plenty convenient to Pittsburgh, a big plot of land lay along the South Fork of the

Conemaugh River. And on that land was a lake. Well, no, not a lake. The makings of one.

More than thirty years earlier, up there above Johnstown, the South Fork had been dammed up. Rocky and fast, the South Fork brought drainage from many other rivers down into the Little Conemaugh, the longer, bending river falling down the valley into Johnstown, where it joined the Stony Creek to become the Conemaugh proper, flowing northwest all the way into the Allegheny River above Pittsburgh. The South Fork had been dammed all those years before because the Little Conemaugh, of which it was a tributary, had once served as a water source for one of the most astonishing projects the state of Pennsylvania, or anyone else, had ever achieved: the Pennsylvania Main Line Canal. That canal system, in its heyday, had needed a feeder reservoir. But the heyday had been surprisingly brief.

Running all the way to and from Philadelphia and Pittsburgh, in competition with New York's Erie Canal, and facing the unique challenge of the Allegheny barrier dividing the eastern and western segments of the state, the Main Line Canal was completed in the spring of 1832. Amazingly, barges could suddenly make the trip between Philadelphia and Pittsburgh in only three days of steady movement, not only pulled by mules on towpaths through canal water but actually hauled, where necessary, over the precipitous mountains by a series of steam lifts and rail lines and inclined planes with counterweights, then returned to the canals to be hauled again by the mules. Nobody had ever seen anything like it: the "portage system," as it was known, combined steam, hemp cables, narrow-gauge rail, and weights, and the English novelist Charles Dickens, for one, took the ride just for the experience of looking down from a canal barge on rooftops and trees. This was a state-funded project, one of the wonders of the age, and incredibly expensive.

Then, in the 1850s, Andrew Carnegie's friends at the Pennsylvania Railroad built the equally daring and justly famous Horseshoe Curve. That was a huge rail embankment project, hand built and hand blasted by Irish laborers. Taking in the vertiginous, panoramic view out over a deep valley, train passengers could see, ahead on the long curve, an oncoming train hugging the hillside or hanging over the air more than a mile away. That curve defeated the mountainous obstruction and got a steam engine from Philadelphia to Pittsburgh even faster than anything the canal system could achieve, and with far greater capacity.

So both the portage process and the whole canal system were abruptly made obsolete. They were sold off by the state to the railway company that had made them so, and as Carnegie and his friends were off and running with rail, the whole amazing canal project went into disuse forever, its parts sitting around doing nothing. The canal had moved fast, the train faster, but fastest of all was the pace of change itself, with the concomitant abrupt discard of systems that had seemed, for a moment, miraculous engines of the future, but quickly turned out to have been only technological byways, money sinkholes, relics whose crumbling infrastructure would hang around sadly while the real world raced on past.

In the meantime, however, the South Fork had been dammed. Back when the amazing canal system was still early in its short life, it became clear that the Conemaugh River, which normally fed the canal, could get too low in summer dry spells. The state therefore needed a feeder: a reservoir, higher up, with sluices to let supplementary water run down into the canal. In 1838, the canal's head engineer explored the region of the South Fork and picked out a good site for a dam that would raise the creek and push it out, along with its many feeders, flooding a four-hundred-acre valley between hillsides into a broad reservoir.

Dam construction didn't get going until 1840, and even then the state quickly ran out of money—the canal system, for one thing, was bankrupting it—and then there was a cholera epidemic. So work on the dam didn't resume in earnest until 1850, and nobody working on it knew that within a few years, thanks to the railroad, all of this effort would be for nothing. The state's contractors and builders therefore threw their hearts, souls, brains, and backs into building the South Fork dam.

Like all other effective dams of the time, the South Fork dam was made of earth: horizontal layers of clay. Each layer was pounded down tight under a standing layer of water: that way, the earthen layer would become waterproof; only then was the next layer added. At that painstaking pace, the structure eventually rose 62 feet high and 930 feet long, a curving form 270 feet wide at its base and only about 20 at its flat top. The nearly perpendicular outer surface was stuck with gigantic rocks, added to make it look like a big work of fieldstone masonry; the more sloping inner surface was similarly coated, but with smaller rocks.

On each of its ends, the dam was attached to the hillsides that rose high above the banks of the South Fork. And at the eastern end, a 72-foot-wide spillway was excavated a good 8 feet below the top of the dam, cut not through the earth core of the dam itself but through the solid hillside rock, using dynamite to blast a channel, dropping quickly to bring water cascading down a series of rock steps through the woods to enter the natural river channel below. This spillway's purpose, like that of any passive drain, was to relieve the reservoir of its water whenever the level should reach the spillway's height, well below the dam's top. The water level, as relieved by the spillway, could thus never climb high enough to run over the top of the dam—"top it," as they said. Water topping an earthen dam and spilling over its outer

71

face will inevitably wear down the face, erode the earthwork, and bring the dam down.

The reservoir resulting from damming the South Fork in that manner was meant to supply the canal below, during drought season, not via the spillway—that was just to control water level—but through five iron pipes, each two feet in diameter, set in a stone culvert running through the dam at the center of its base: the canal would thus be fed from the bottom of the reservoir. A wooden tower, connected to the dam and set above the lake, near the inner face of the dam, held controls for opening and closing valves in each of the five sluice pipes, thus precisely regulating the discharge, holding water in the reservoir or letting it out, in a finely controlled manner, down into the canal.

In June 1852, the reservoir's sluice pipes were closed up and the South Fork began flooding the whole clear-cut, four-hundred-acre area, in order to feed the soon-to-be-obsolete canal. It took about three months for the river, thus blocked, to rise and spread itself and flood the huge, broad area: a reservoir, ultimately about sixty feet deep, two miles long, behind the South Fork dam. In the fall, water from the reservoir began feeding the canal via the sluice pipes operated by the controls in the wooden tower. This was only about a year before the first steam engine was to run the rails from Philadelphia, round the stunning Horseshoe Curve, roll on to Pittsburgh, and make the whole canal project, along with the reservoir and the dam, irrelevant.

That was a very good dam. Had it been maintained, it would have held up for decades and survived extreme flood conditions.

But it wasn't maintained. The Pennsylvania Railroad, having superseded the entire canal system, bought the whole canal system at a bargain price from the state, with a view to privatizing it. There wasn't much percentage in that, though, and what the company really did is ignore the parts of the system that didn't involve rail. The system's many nonrail structures—tunnels with stone archways, inclined planes, pulleys—collapsed to ruin in the woods, more romantic and mysterious as the years went by. The company did post a watchman at the reservoir it had bought along with everything else. But otherwise it ignored the dam.

Posting a watchman was a good idea: in its state of neglect, the dam broke in 1862 during a record downpour. That's because the culvert with the sluice pipes had degraded, and the dam sank and partly collapsed around that point at the center of the base. Happily, the watchman had opened the pipes during the storm, letting off a lot of water, and the abandoned reservoir was only half-full anyway when the breach and sinking occurred. While the sudden, uncontrolled spill did some damage to the railroad's tracks, it did little harm to the towns below.

The dam itself, however, so lovingly and expertly con-structed, had now partially collapsed. Even while it had re-mained in place, it had lost integrity; it sagged and leaked and after a while almost all of the water had leaked out. There was no reservoir any more, just a big, ten-foot-deep puddle and a wide grassy area where water had dried and vegetation revived. Sheep and cattle were sometimes left to graze there. At some point, the wooden control tower burned to the ground, and by the 1870s the whole thing was nothing but a weird, wet, woodland ruin.

Until, that is, the old collapsed dam and wetland, having served so briefly as the canal system's reservoir, caught the eye of Benjamin Franklin Ruff.

The Adirondack lakes, on which the eastern establishment elites built their gigantic woodland "cottages," lay 1,500 feet and more above sea level, where the air was fresh and cool. Lake Placid, for example, one of the most beautiful and popular of those Adirondack retreats, was built on a spot nearly 2,000 feet high. Yet around those lakes are mountain peaks: Whiteface rises nearly another 3,000 feet above Lake Placid, Mount Marcy looms 3,000 feet over Elk Lake. That vast New York wilderness is filled with high, broad, rolling valleys and bowls where glaciers left nice lakes.

The Alleghenies, lower than the Adirondacks, consist of a series of parallel, curving ridges, each rising with impressive steepness out of the valleys, not featuring any sweeping bowls or long alpine valleys. The beauty of the Alleghenies lies not in towering heights and peaks but in ruggedness: you're either climbing or descending, hard; there are outcroppings and drop-offs and vertiginous views. The glaciers that left standing bodies of water behind moved across only a small portion of Pennsylvania, leaving the rest alone. Nature thus allows for no big lakes at all up in the Alleghenies. There's nowhere there for spring and river water to collect and flush and refill.

But Benjamin Ruff had a vision, and the resort, he determined, could be built up high, not far from the crest, where the air was the nicest, just like Carnegie's beloved Cresson, because there was already a big spread of timber-cut land, overgrown now but by no means reforested, and there was even a crumbling ruin of a dam that could be easily rebuilt, blocking up the South Fork and creating the kind of sparkling mountaintop lake that invites fishing, boating, recreation, and restoration. By the time Ruff was looking to buy the property, in 1879, this place with the ruin of a dam no longer belonged to the railroad company but to a one-term congressman, John Reilly, serving in the House on hiatus between long terms as an official of the railroad. He'd bought the property from his employers there for $2,500 in 1875. Reilly did nothing to develop the property, and when Ruff came along, the congressman was so happy to unload it that he took a $500 loss in the sale.

Now in possession, Ruff created twenty-one shares in the proposed venture, set aside four for himself, and offered the rest, at $200 per share, to fifteen other big men of Pittsburgh. When fourteen of them bought a share each, at $200, Ruff might have seemed to turn a profit, but there was money to be spent on transforming the property. Membership in the club was to cost $800, and to keep things exclusive, no more than one hundred members were to be admitted, along with their families, at any given time. There would be a clubhouse, with rooms to let, and where the members and their guests would be served meals, but in order to encourage the members to build lakeside cottages, no stay at the clubhouse could exceed two weeks.

The fifteenth shareholder bought three shares. This was Henry Clay Frick, only twenty-nine at the time, yet already the world's most famous coal baron and a close friend of Benjamin Ruff. When the club formed, Andrew Carnegie wasn't yet a member, and Henry Frick was the biggest name on the secret membership rolls. As a founder, and as a controlling member, Frick would have much to do with decisions made by the club and with what happened at the end of May 1889. But nobody would ever find out exactly what that was, because Henry Clay Frick kept everything to himself.

His mother, Elizabeth Overholt, was the child of wealthy Mennonites of West Overton, near Pittsburgh, immigrants from Germany, owners of much property around West Overton, including an entire village, as well as two big, commercial distilleries that made the rye whiskey later to be branded as Old Overholt. These weren't artisanal whiskey producers like those who had become whiskey rebels, but commercial beneficiaries of the federal suppression of the rebellion. One of the Overholt plants could put out 860 gallons per day.

Yet Henry Frick's father, John, was poor, a farmer barely making it. Henry grew up with a fervent desire to emulate his rich Overholt grandfather.

And by 1879, not yet thirty when he and Ruff took a controlling interest in the South Fork Club, Frick had well surpassed that goal. Relying more on charm and drive than good credit, in 1871 he got loans from the Pittsburgh banking family the Mellons—he was good friends with young Andrew Mellon—and went into business in the Connellsville, Pennsylvania, coal fields. He soon began supplying the iron business and then the steel boom with what its furnaces needed every hour of every day: coke. Having just begun that business, Frick saw that the financial panic of 1873, driving investment value down everywhere, could benefit him. Starting at 3:00 A.M., he would supervise his coke ovens in Connellsville, then look for deals on failing properties, then go to Pittsburgh to take orders for coke, then head back to Connellsville to do the daily books. In that way he began to corner the coke market. When a feeder railroad was about to go under, leaving Frick with no way to get his coke to market, he bought the controlling shares in the rail company at rock-bottom prices. That gave him ownership of his means of moving his coke. Risking heavy buying in a down market, and believing that coke must rebound, by 1878 Frick had one thousand ovens producing one hundred freight-car loads of coke every day, and he was right: coke had rebounded. He was a millionaire.

Still ahead of him: his famous union-busting, collaboration, and competition with Andrew Carnegie, and acquisition of European art on a mass scale. Yet Frick already wanted to get away from the smell of the coal his workers labored to mine and burn. He needed restoration in the mountain air in the company of his wealthy peers. Frick encouraged Ruff to buy the site of the old reservoir and form the club.

With seven shares of the South Fork Fishing and Hunting Club between them, Frick and Ruff formed a strong majority combination. Ruff wrote a corporate charter to form the club, registering it with the Allegheny County Court in Pittsburgh, where the organization's offices were to be maintained, and not in Cambria County, where the property was located. That was illegal, but Judge Edwin H. Stowe of Pittsburgh, a veteran jurist beloved and endorsed for years by all parties and factions of influential men, had no problem overlooking the requirement that corporate charters be entered with the court of the county in which they operate. The people and officials of Cambria County thus had no warning of what was about to go on, up on the mountain above the industrial towns along the Conemaugh River.

The Ruff-Frick majority shareholder combination also took the lead in creating the club not only on paper but also physically. The big thing, of course, was making a lake. That meant redamming the South Fork Creek.

The question facing Ruff was how to get the old dam back, as inexpensively and quickly as possible. In the fall of 1879, the club had more than sixty members eager to hunt, fish, boat, and relax, out of the heat and stink of Pittsburgh. They wanted to get that going next summer. Ruff's first idea was to rebuild the dam to a height of only forty feet—twenty feet shorter than the original—while compensating by cutting the spillway twenty feet deeper. That sounded cheaper and faster at first, but when a closer look suggested otherwise, Ruff bit the bullet. The dam had to be repaired and rebuilt, if not to its original height, then nearly so. He'd overseen the digging of railroad tunnels. He had no doubt he could engineer dam repair.

The first step was to disable and fill in the stone culvert in the center of the old dam's base, where the discharge sluice pipes had run. That's what the dam had collapsed around, when it broke in 1862. But to do that, Ruff needed labor, and that's how the people of Johnstown and the surrounding towns first gained confirmation that a massive construction project was going on, nearly five hundred feet over their heads. On October 15, 1879, an ad ran in the *Johnstown Tribune* offering work to fifty men. The organization proposing this work wasn't named, but of course the location had to be revealed to the workers, and rumors had already flown that some sort of club was building a summer resort up there. Now it was out in the open.

Membership in the club, however, was not. Nor was any club charter on file in the Cambria County courthouse.

Along with the workers, service people carting materials gained access, too, as well as the unavoidable bystanders. So as work began on stabilizing the old culvert at the base, with the purpose of shoring up the derelict dam, residents of the valley below, and of the little town of South Fork, not far from the club's property, began to see just what Benjamin Ruff's ideas about dam engineering really involved. Stories trickled down of what was going on up on the mountain.

These accounts conflicted. Some said the five rusted sluice pipes had already been removed: Reilly, the former owner, had sold them for scrap. Others were sure the pipes were still there, in degraded form: Ruff's process involved encapsulating them with layers of hemlock boughs and other material. Everybody agreed on one thing: Ruff's first big move in rebuilding the dam was to dump and pack anything and everything that could be found to fill up the old stone culvert at the bottom far below the steep hillsides. Workers brought rocks from the hillsides, hemlock from the woods, and hay and manure from the farms. That all got packed into the culvert. Whether the pipes were long gone or simply covered up and blocked, the culvert was clearly no longer operational. There was to be no water discharge, controllable by any mechanisms, via the dam itself. Only the passive spillway would relieve high water.

The main earthen structure, meanwhile, got no significant repair, despite the collapse's having lowered it and let water get into it. The earth structure was raised, but in the end not quite to the old level.

And so the dam promptly failed. With work ongoing, on Christmas Day of 1879 a downpour washed away what had been completed. Nothing could resume until the summer of 1880, postponing sport and relaxation by a year. On Ruff's engineering model the dam was more or less completed in January 1881. Then a February rain caused further damage. Repair work began.

Still, after all that, by the end of March 1881 there was an actual rebuilt dam, and it was an amazing and impressive thing to see. Viewed from below, the earthwork loomed above the valley of the Little Conemaugh like a work of the gods. Its flat top was more than twenty feet wide and more than seven hundred feet long, connecting the high hillsides. For the club members' comfort, a carriage road traversed the chasm on the dam's broad top, offering a commanding view out over the whole countryside that fell away far below. On the spillway end, two trestle bridges crossed above the cascade to connect the road on the hill to the road atop the dam, one bridge for each traffic direction. Also by March, the South Fork had swelled and risen behind the dam, flooding the whole area and becoming a clear, broad lake more than two miles long—in spring floods it could be three miles—and sixty feet deep.

That's what they'd really been building, of course: not a dam but a lovely lake. It was time to get crews started building small steamboats, and to stock the lake with the fish sportsmen coveted. One thousand black bass, bought at one dollar per fish, traveled by rail from Lake Erie in their own tank car. Only one died along the way. Up the mountain they came, by horse-drawn wagon, to be released in the deep water of what truly now was a lake, filled not only with imported bass but also with the trout that came naturally with the South Fork Creek.

Just as naturally, the fish would want to leave the scene, via the spillway cascade, whenever the water rose high enough to usher them out that way. But Ruff had thought of that. To keep the prize bass in the lake, he ordered a complicated apparatus installed at the spillway. This fishguard, as it was called, involved hanging a row of iron rods from the trestle bridge nearer the dam's inner wall. The rods plunged into the water of the lake before the spillway and formed, with wire netting attached, a barrier to exit. Further discouragement came from a boxlike structure made of logs bristling with nails, forming a V with the point facing into the lake, floating vertically on the water's surface, rising and falling with the lake's level to guard the spillway, and also secured to the trestle's posts. The whole apparatus would have struck a more experienced dam engineer as more or less designed to defeat, or at least seriously compromise, the tendency of a spillway to spill. But the fish weren't going anywhere.

Ruff, Frick, and the other members had put up with a lot of delay, thanks to washouts. Now they could look with satisfaction upon their wide, glistening mountain lake filled with fish leaping for the hook and little steamboats plying the waters: a lake ready for sailing and fishing and the building of lakeside homes. They named their new-made creation Lake Conemaugh, and in the summer of 1881, relaxation and recreation began in earnest at the South Fork Fishing and Hunting Club, to the great dismay and disapproval of Daniel J. Morrell.

CHAPTER 4

"NO DANGER FROM OUR ENTERPRISE"

WHEN LOCAL CREWS BEGAN IMPLEMENTING BENJAMIN RUFF'S approach to dam engineering, in the fall of 1879, Daniel Morrell decided he didn't like what Ruff was up to, didn't trust Ruff's engineering instincts, and was going to make Ruff alter the dam project to ensure the safety of the people and property—the Cambria Works, the Gautier wire factory, the whole city, and everything else Morrell owned and for which he took responsibility—down below.

The club membership was secret, of course. Yet ordinary people down in the valley and back in Pittsburgh could guess who might have joined, and from the time the club began operations in 1881, and then as the membership swelled toward its maximum enrollment throughout the decade, and right through the disastrous events of May 31, 1889, people did guess.

"The Bosses' Club": that's what an article in a Pittsburgh

paper called the South Fork Fishing and Hunting Club. And everyone knew who the biggest bosses were: Carnegie and Frick, of course, in steel, coal, and everything else; Andrew Mellon the banking scion; and Robert Pitcairn, head of the western section of the Pennsylvania Railroad. These were the first men of Pittsburgh, a status that placed some of them among the leading men of the nation.

The mystery of club membership only amplified anxiety. People living down in Johnstown and the other towns below the newly created mountaintop lake had obvious reason for concern. News from the mountain of Ruff's engineering style in rebuilding the dam, the failure of the disused canal dam back in 1862, and the washouts even while reconstruction was going on made people nervous, and they wanted to know for sure who was behind the project. It was hard to find out anything much about the South Fork Fishing and Hunting Club. The club's management, as deployed by Benjamin Ruff, made it as hard as possible for any ordinary person to wander onto the grounds and poke about. Still, as the grand, brightly painted Queen Anne cottages got built and landscaped, and the white sails flew across the lake under crisp blue skies, and the well-turned-out big-city families arrived by train to be taxied, along with their servants and trunks, up to the lake by horse-drawn spring wagons and buggies, the labor, skills, and services of plenty of local people from the Johnstown area were needed at the club—there were lawns to manicure and flower beds to tend, food to serve, and boats and a clubhouse to maintain—a lot of local people got a good sense of what the place was like, and of the sort of people who frolicked and rested up there.

So a fair number of people got a good, hard look at the dam. They weren't blind. They could imagine—they thought they could, anyway—what might befall the towns below should the dam break and the water in the lake come down the mountain all at once.

What they didn't have was any good way of expressing their discomfort and getting something done about it. There were no laws or regulations regarding damming up a property; there was no government charged with oversight; there was no collective clout that the worried citizens of the valley could bring to bear. Even if there had been, the city of Johnstown was divided into many separate boroughs, each with its own independent government, and no love was lost among them. There was simply no regular political process for expressing the people's will, no law to which anybody could appeal. Conflicts like this came down to a matter of leading-citizen types undertaking responsibility for pressuring other leading-citizen types to do what was right.

In the matter of the South Fork dam, responsibility for addressing concerns about the project and doing something about it lay with the one man whose identification with both the Cambria company and the city of Johnstown was total: Daniel Morrell. Not only the head of the bank, the gas- and waterworks, and intermittently the city council, Morrell had by now also served two terms in Congress and had a lot of pull in the state's Republican Party. These were his towns, these his people, and he liked to think that while he busted unions without compunction, he took care of his people. Morrell also had plenty of reason to suspect the glittering Pittsburgh crew of heedlessness, even flightiness, when it came to practical matters like dam construction. His interactions with Carnegie had underscored this feeling; recently he'd also had some conflicts with Robert Pitcairn, western head of the Pennsylvania Railroad and—though nobody outside the club knew it for sure—a member of the club.

Morrell was never against new development. He didn't wish

to prevent the club from making a lake. But he was opposed to anything carried out in a slipshod manner, and he'd weathered enough fires and other accidents at the Cambria Works and other job sites to know the degree of diligence required to prevent disaster. Nobody in Johnstown had a bigger foot, and Morrell was determined to put his down and get some satisfaction from the Pittsburgh sportsmen on the mountain.

When John Fulton, Morrell's top mining engineer, arrived at the abandoned site of the old reservoir to observe what was going on, Benjamin Ruff's hired crews were still in the process of working on the dam. John Fulton was probably the most qualified and experienced engineer in the Johnstown region and possibly well beyond: he was a geologist, too, and a Bible-shaking Presbyterian temperance activist, with no regard for rank or wealth. To the club officials, Fulton's errand of inspection was far from welcome, but they couldn't easily deny a polite request by Daniel J. Morrell of Johnstown and the Cambria Works for an inspection by his top engineer. So a delegation of the Pittsburgh sportsmen met Fulton at the site to give him the tour.

These men included the club's president, E. J. Unger, a railroad man and, in partnership with Carnegie, an upmarket hotelier; and R. C. Carpenter, owner of Crawford Coke and Coal, like Henry Clay Frick an extractor in the coke fields of Connellsville. Also on hand for the clubmen was their own expert, N. M. McDowell, a civil engineer of Pittsburgh.

Escorted by these gentlemen, John Fulton turned his cold and practiced eye on their dam repair project. He saw right away what was wrong.

There was only about forty feet of water in the lake now; the projection was to get sixty, and that's about how deep the old canal company reservoir had been, too. So the size and heft of the dam should be enough, Fulton figured, to hold back that immense tonnage of water. And at seventy feet wide the old spillway was ample.

But Fulton watched with amazement at the workers filling in the old culvert below the breach that had caused the partial collapse that had emptied the reservoir. That breach was being not repaired but only obscured by this process of filling in the culvert. The breach still existed within the dam. Once the lake had flooded in and risen behind the dam, finding the breach and fixing it would become impossible.

Which raised the most important problem. With the old sluice pipes gone, or made deliberately inoperable, there was no active way to lower the level of the lake. The spillway, cut eight feet below the height of the top of the dam, could lower water only passively: in the normal course of a rise of water, the level wouldn't go higher than the spillway. But in a major rainstorm, with water rushing into the lake from the South Fork and all of its many tributaries, that method of release might be quickly overwhelmed. If water rushed in faster than it could be released by the spillway cut, the level would rise toward the top of the dam, slowed but not stopped by the spillway, and if it kept raining, eventually water would top the dam. If that happened, sooner or later the dam would go, and the whole lake—20 million tons—would fall down the valley on the towns below.

Even in the absence of a major storm's swelling the South Fork, should even a minor leak occur, presaging future trouble, there would be no method to lower the lake in a controlled way, releasing water carefully into the Conemaugh below, make an inspection, and repair damage. There was no deliberate way to get water out of the lake.

◆ ◆ ◆

So just as Daniel Morrell might have predicted, the rich Pittsburghers' design violated in every imaginable way the most basic premises of engineering a dam for safety. It was literally only a matter of time, Fulton judged, before the breach caused flooding.

How much flooding? That was impossible to say: it would depend on the height of the water and the size of the breach. How disastrous a flood? Unknown. But disastrous enough, no doubt.

There was only one way to prevent this otherwise inevitable outcome. Along with fully overhauling the rocky outer coating on the upper part of the dam—that too had been neglected, Fulton judged—a discharge pipe system must be designed and installed to allow for deliberate reduction of the water level. And the breach around the old culvert must be fully repaired.

Fulton reported all this to Morrell. Pulling no punches, he underlined the phrase *only a question of time*. Morrell, disturbed but hardly surprised, forwarded Fulton's report to Benjamin Ruff in the latter's capacities as club founder and supervisor of the dam project. Morrell couldn't have expected Ruff to express gratitude for this negative report. The necessary repairs would obviously be costly and time-consuming. But given the stakes involved, Morrell might have expected grudging compliance, at least.

Ruff wrote back, opening his letter with pro forma politeness, then dismissing out of hand each of Fulton's criticisms. For the first time, Ruff revealed the name of the club—but only because Fulton's report had gotten it wrong, calling it the Sportsmen's Association of Western Pennsylvania, and Ruff made that error his first point of dispute. He was implying that if this man Fulton got the name of the club wrong, he couldn't be very punctilious in general. That was just for starters.

"In the second place," Ruff went on, "he is wrong in saying that the dam was originally built of stone." That wasn't what Fulton had meant, but Ruff also denied that the inner face had ever been lined with stone, and Ruff was wrong about that.

Fulton had in fact made a few minor errors. He didn't know about the wooden operating tower, exactly how many pipes had been in the old culvert, or exactly the size of the old breach that Ruff had supposedly repaired. From those mistakes, Ruff drew the conclusion that Fulton's proposals for repair had "no more value than his other assertions." Ruff simply ignored the idea that the lake would need some means of being actively emptied or lowered, and signed off with a tone of not expecting to be bothered about the matter again. He attached a report from the club's paid engineer, McDowell, asserting that the dam was safe.

Morrell, now deeply concerned, pushed back. In a letter to Ruff, he admitted that Fulton might of course have made some errors but noted that the main issue remained impossible to dismiss. Without some means of actively lowering the water, the dam couldn't be properly secured.

"We must protest," Morrell told Ruff, "against the erection of a dam at that place, that will be a perpetual menace to the lives and property of those residing in this upper valley of the Conemaugh, from its insecure construction." He enclosed a further report from Fulton, stating that other independent engineers had since looked at both Fulton's first report, and at the dam itself, and agreed with Fulton. And he politely sent back McDowell's report, with the comment that the club's engineer might want to rethink it, in light of these issues. Morrell went further. So seriously did he take the situation that he offered Ruff a contribution of Cambria Works money and expertise in a joint project to secure the dam.

Ruff must have scoffed in irritation: why would the club

need help or money from Daniel Morrell? In any event, that offer was ignored, and the dam work kept going on as planned.

Morrell saw that he was being decisively locked out of the dam process. So he took the next step. He joined the South Fork Club. Buying two memberships in his own name, he got himself on the inside of the problem. This way, he could go personally to the lake more or less at will, keep an eye on the dam, and agitate and network within the membership for safety improvements over time.

As the summer resort began to come to life in the early 1880s, it started taking on just the mood of elegant relaxation Ruff and Frick had hoped for. Members built generously proportioned summer houses along the lakeshore, with gables and peaks and other impressive features. A system of boardwalks connected the cottages to one another and to the big Stick-style clubhouse with its deep porch. Some of the cottages had docks for swimming, fishing, and tying up boats.

Life at the club was leisurely and lively. The families slept in their cottages but dined together in the big clubhouse dining room, which featured a twelve-foot-high brick fireplace faced with decorative tile. They wore fashionable summer clothing, and bands were hired to come up and play for the guests. There were impromptu amateur musicales and theatricals, and elegant picnics by the spillway: cascading over rocky steps, it gave a ro- mantic, wilderness-waterfall effect. And of course they hunted and fished, and sailed, steamboated, and paddleboated. The top of the dam with its amazing view and horse-and-carriage road was always an attraction.

As the 1880s went on, the leading club members also solid- ified and extended their positions as great men of business. It was just after getting the club going with Ruff that Henry Clay Frick's enterprises really began taking off: Frick rose throughout that decade to become second only to Carnegie in the ranks of the Pittsburgh industrialists. Frick made key deals with Carnegie through which each man verticalized his enterprises via one another, guaranteeing Carnegie as much ready coal as he could use, at a good price, with no competition, and Frick a market for coal so stable that he was effectively already in the steel business with Carnegie. Soon not only did the H. C. Frick Coke Company employ eleven thousand men, with $5 million in capital and thirty-five thousand acres in coal mining and development, but Frick also became chairman of all of Carnegie's world-dominating iron and steel efforts.

The club prospered along with its members. The secret membership roll grew quickly, studded with the biggest names in Pittsburgh high society. What those rolls would have revealed, if anyone had seen them, was the members' almost endless interrelatedness, and in particular, their mutual connection to one man, Andrew Carnegie. Big Pittsburgh manufacturers, bankers, lawyers, and entrepreneurs who were among the club's members included Robert Pitcairn, one of the most powerful men in the region. He'd replaced Carnegie in running the western section of the Pennsylvania Railroad. Like Carnegie, he was Scottish-born, and like Carnegie he'd worked his way up quickly in the railroad company. Now his rail yard along Turtle Creek near Pittsburgh was the largest in the world. Pitcairn and Carnegie were friends, and Carnegie of course always maintained the tightest kind of alliance with the railroad company.

The club member Henry Phipps, a financial wiz and long-standing money partner of Carnegie, was now chairman of the cluster of companies known as Carnegie Brothers & Company and Carnegie, Phipps & Company. John G. A. Leishman, a club member, was a Carnegie executive, and James Hay Reed and Philander Knox, partners in their firm Knox & Reed, gave Carnegie his two top lawyers for both corporate and personal matters; they were members. The member Elias Unger was a partner with Carnegie in Pittsburgh's Seventh Avenue Hotel.

And then there was Andrew Mellon, the best-known member of the club, along with Carnegie and Frick. Mellon began in the family banking business and would one day have powerful interests in every kind of big industry. He had backed Frick financially in the early days, and now the two were close, traveling to Europe together to look at and snap up great works of art on an astonishing scale.

The club's list, that is, distilled the simple fact that not only steel but all of large-scale American production in the region had become concentrated in the hands of a very few men. They were all related in one way or another to Andrew Carnegie, Henry Frick, and Andrew Mellon, and to one another.

Meanwhile, down at the bottom, life went on, too. Over the years, people in Johnstown and the other towns got used to the presence of the fancy families up on the mountain. Things were always tense, but the friction of inequality too became part of life, going on, as it does.

On the bright side, the club meant employment, for some: press about the beautiful location in the Alleghenies suggested other resorts might be built, too, with yet more benefit to the people of Johnstown. Also on the bright side, boys and men alike could, at first, sneak easily onto the grounds, ignoring a multitude of no-trespassing signs, and catch trout in the streams and bass in the lake. But soon the club's management fenced off the best trout streams. When locals removed the fences, the club officially announced an intention to shoot on sight any nonmember on the grounds at night. That didn't make the bosses' club any more popular with the locals. But when it came to the dam, while people didn't exactly stop worrying, worry about the dam became just another part of life in Johnstown.

One factor in the city's grudging acceptance of the dam was a false alarm. It occurred early in the life of the club's lake. In June 1881, with the club just up and running, there was a flash flood of the Conemaugh, and suddenly Johnstown was alive with the idea—spread by word of mouth—that the dam up at the lake was on the verge of breaking.

From the little hillside town of South Fork, up near the club, down to the streets of Johnstown proper, word flew that the inevitable was about to occur, and Daniel Morrell, alerted to rumors of what he'd already predicted, sent two men from the Cambria Works up to the dam to get a look. By the time Morrell's men got there, the water level had risen to nearly two feet below the top of the dam. The spillway was splashing and crashing, and just as Fulton had predicted, the fastest possible spill through that cut just wasn't anywhere near fast enough to bring the water level down to the spillway's height. Still, the rising had slowed; there was no leakage; the spillway was operating. Morrell's men came back to Johnstown with that reassuring report. It didn't help: people in town were in a state of intense anxiety, and at the western, lower end they stayed up all night in the pouring rain in sheer terror.

But then the rain stopped and the sun came out. The inspectors were right. The dam had done fine.

A false alarm like that can make any concern seem silly. Flooding came nearly every spring, and sometimes even in the fall. Now every time the streets flooded people would make comments and jokes about the dam breaking. Every time, the dam held. Over the years the jokes and remarks, reflecting collective anxiety, kept outright panic at bay.

For even as the urgency about repairing the bosses' dam subsided, flooding itself was getting worse. The erosion of the stripped mountainsides was combining with the narrowing of the Little Conemaugh and the Stony Creek, by the industrial waste and the building and rail landfill, to limit drainage and make flooding common. Town ordinances had even taken note of the issue. The width of the Stony Creek had now been set by law at 175 feet, that of the Little Conemaugh at 110. And yet the Conemaugh proper, receiving water from those two rivers, with a total legal width of 285 feet, was itself now only 200 feet wide near the mills.

Anyone could do the math on that. And nature could do it for you. In a storm in 1875, the Conemaugh rose two feet in an hour. Eighteen eighty saw the breaking of a dam on the Conemaugh, built by the Cambria Works, though in a relatively safe location below town. Then, from 1885 to 1888, came three bad floods, with the Stony Creek rising three feet in forty-five minutes in 1885. In '87, in the lower end of town, not only basements but ground floors were flooded, about a foot high.

By May 1889, some of the town's leaders were talking about requiring the Cambria company to take cleanup and demolition measures to bring the rivers back to natural width and depth. Yet even as the flooding trend was clearly worsening, public anxiety about the dam up at the lake was not. James Quinn was in the minority: "too fearful" about the dam, as his sister-in-law Abbie said. The *Johnstown Tribune* took the doubly reassuring position that a) the club's dam was clearly holding up fine; and b) even if the dam broke, water from the lake would have plenty of room to spread out and dissipate throughout the valley before it could do any significant damage to the Johnstown region below.

Only the big, gruff Daniel Morrell of the Cambria Works had directly challenged Benjamin Ruff's engineering, as well as Ruff's blithe certainty that the project exposed Johnstown to no danger. Having gotten nowhere, Morrell had then joined the South Fork Club for the express purpose of monitoring the dam and working from within to get something done about it.

Yet right after joining the club, Morrell suffered at first a slow and finally a total collapse of his own. Having lost business to Andrew Carnegie's juggernaut, Morrell also lost his grip mentally. By 1884, he let go of all of his responsibilities, both at the Cambria Works and in the city of Johnstown, and went into seclusion in his big house on Main Street. He died there in August 1885. Two years later, Morrell's opponent on behalf of the club, Benjamin Ruff, died too, of an infection.

So neither man would know the events of May 1889, which would prove beyond all doubt which of them had been right about the quality of the dam.

There was another man with clout in Johnstown who harbored suspicions that the dam was no good. Tom L. Johnson was thirty-five in 1889, and though he didn't live in Johnstown, he had substantial business interests there. He also had vaulting ambition—he'd begun his career in hopes of becoming another Carnegie—accompanied by an unusually frank sense of humor about the ruthlessness of the business world. Where other men used terms like "capturing markets," Tom Johnson unabashedly identified himself as a monopolist. He was equally clear-eyed about the kind of chicanery that went into a monopolist's success.

Tom had started out poor, moving about the South with his family after his father was ruined in the cotton business. At the age of eleven, he saw his future. Hanging around the railroad station in Staunton, Virginia, where his family was then living, the boy became friendly with a conductor, who offered to bring newspapers on his train for Tom to sell at the station; Tom got the conductor to agree not to bring papers for anyone else, so Tom could set prices, with no competition. Young Tom made easy money selling papers and saw right away the incalculable value of private supply deals—and of being the only game in town.

Eventually the deal ended when the conductor was transferred. But Tom had decided what his business was to be. Not newspapers, not trains: monopoly.

Seeking out businesses in which there was little or no competition, in 1869 Johnson began working with the du Pont brothers, famous capitalists, in a streetcar business they'd acquired in Louisville, Kentucky. Tom patented some minor streetcar improvements, but more important, he learned how getting ahead in the burgeoning business of public conveniences like streetcars had more to do with political cronyism and influencing civic policy, and less with knowing anything special about transportation. This insight, he would later say with relish, made him a nationwide supposed expert in streetcars, and soon Tom owned controlling shares in the streetcar systems of Indianapolis, Cleveland, St. Louis, Brooklyn, and Detroit. He moved to Cleveland in 1883 and bought a mansion on Euclid Avenue, the stretch known as Millionaires' Row.

Meanwhile, having invented—or, as he said later, pretending he'd invented—a peculiar kind of streetcar rail, and protecting it with what he later noted was a weak patent, in 1883 he and his partner Arthur Moxham made a deal with Daniel Morrell at the Cambria company to roll these special new rails in Johnstown. That's what brought Tom Johnson to town.

Soon, with Cambria cooperation, Johnson and Moxham built their own rolling mill, in a town they immodestly named Moxham, just south and uphill from Johnstown. They built their own steam railroad to run from Moxham into Johnstown. With Daniel Morrell succumbing to dementia, and giving up his offices, Johnson naturally also hoped to get control of the city's streetcar system. By late May 1889, Tom L. Johnson's new, independent steel mill in Moxham was all set to start rolling steel rails.

Yet a few things had been happening lately to give Tom a new view of life. He was qualifying the only real interest he'd had for so long, the pursuit of monopoly.

He read two books. Both were by the political philosopher Henry George: *Social Problems* and *Progress and Poverty*. George argued that the horrors of poverty flowed directly from industrialists protecting their new wealth from taxation by keeping it locked up in land with rising value. George wanted a tax on land, not production. The idea was to level a certain amount of privilege by releasing some of the wealth of the privileged for the good of all.

Reading Henry George, something occurred to Johnson. So far, his interest in privilege had lain solely in getting it. But George's description of what was happening to the poor as a result of wealth rang true to Johnson. He knew the system. It had made him rich. Maybe Henry George had figured the system out. Maybe everything Tom loved was destroying the lives of others.

Tom didn't want to believe that. He took the George books

to his lawyer in Cleveland. It was the lawyer who had made him a convinced free-marketer, Tom reminded the man, and he begged the lawyer to point out all of George's errors. Otherwise Tom felt himself in terrible danger of starting to advocate for George's tax plan. That would make him a traitor to his class. But the lawyer didn't take the demand seriously.

So Tom had already begun questioning both the privilege and the philanthropy of the monopolists whose ranks he'd once been so proud to join, when in the summer of 1888 he too went up the mountain to the South Fork Club to look at the dam. He had admiration for the way Daniel Morrell did business, and he'd heard some bad things about the dam.

Standing there, observing the dam and the lake, Tom felt deep misgivings about the lake's stability. He was no engineer. He thought instead about the privilege so entirely embodied here, a gorgeous mountain retreat that monopoly had built. He speculated on what might happen to himself, his friends, his business, and everybody else down the valley, should that dam ever give way.

Tom L. Johnson was having suspicions about monopoly and privilege. What happened in Johnstown on May 31, 1889, would change the purpose of his life.

CHAPTER 5

RAIN

When the rain started falling on May 30, 1889, it had already been a weird weather year.

Late that winter a series of heavy snows had fallen on the Alleghenies. Then in early April came the heaviest snowfall of the year, eight inches in Johnstown and up to two feet in the mountains above. Due to the snowmelt, runoff alone was unusually high that year; then the whole state of Pennsylvania began getting so much rain that by May 1889 it was on track for a rainfall record for the entire century. As May began, heavy storms banged the whole state hard, with sudden local drizzles flashing on and off with no warning. There had already been ten days of rain that month in Johnstown, and still, up on the mountains, light frost overnight. Temperatures swung wildly, hot one day, cold the next.

Then, on May 26, all the way out in California, a major cyclonic disturbance began moving eastward. As that storm traveled, it made unusually deliberate progress. Given the extremity of the rains it released, the system was moving from the

Pacific Coast toward the Atlantic at a strangely steady, regular pace. As the storm came across the country, it encountered other extremes: in some places, temperatures already high, in others, frost. Facing all that variation, the storm grew even more violent, yet for all the tempestuous precipitation it threw down on the continent, and the circling winds that kept picking up its rotational speed, as a system it kept moving slowly, inexorably, refusing to get wafted away, turn, or break up.

On May 28, Nebraska saw intense rainstorms, and in Kansas tornado-level wind knocked down farms and killed a number of people. The next day, severe rain and winds spread all the way across the Midwest from Michigan to Tennessee. Train travel stopped dead.

One of the stranger features of this unusual weather pattern: for all of the gradual progress the system was making from west to east, those encountering it on the local level were greeted with erratic violence. When the storm arrived in western Pennsylvania on May 30, the U.S. Weather Bureau was predicting threatening weather everywhere in the region. One wouldn't experience the storm as slow, steady, saturating: that's how the system was moving, high above, but on the ground the feeling was wild, spasmodic, even random.

A few inches of rain fell in Pittsburgh, for example, while up in the mountains, where the ground was already saturated with late runoff and earlier flooding, the rain was so hard and so powerful, and it went on so long, that the storm would be counted as the worst rainfall ever then recorded in western Pennsylvania, and maybe the most rain of the century over an area of that size.

That's because another strange thing was going on. There were actually not one but three areas of low barometric pressure—low pressure always encourages storms—moving into western Pennsylvania at the same time. The big system that had drenched, flooded, and even killed its way from California was of course flowing, as so many big storms do, from west to east. But another storm was now moving in from the Southwest. Yet another came from the Southeast.

And there was nowhere for any of these storms to go. For a high-pressure zone, bringing lovely weather all the way along the Atlantic seaboard from Bermuda to Newfoundland, had blocked the way. Low-pressure zones, for all of their stormy violence, bounce off high-pressure zones.

These three storms were coming together as one system, triple-sized now, badly stalled at the top of the Alleghenies. Flood conditions would soon stretch down into West Virginia, and then would reach as far southeast as Washington, D.C., and Richmond, Virginia. Yet it was just below the center of this atmospheric event, of an intensity rarely seen in North America, where the Stony Creek and the Little Conemaugh River met at the headwaters of the Conemaugh. That was the Johnstown floodplain. Just above that spot spread the big, deep lake that Henry Frick and Benjamin Ruff and their colleagues had brought into being behind a looming dam.

Nobody on the ground in Johnstown knew exactly what was going on in the atmosphere, of course. They were expecting rain, possibly heavy. Thursday, May 30, 1889, was Decoration Day, as it was often then called: citizens memorialized fallen veterans by placing flowers on their tombstones. It dawned cloudy. Both Weather Bureau reports and casual observation could tell you it was going to rain on that parade.

But unusual showers had been falling throughout May, and flooding could be expected sooner or later this season anyway. People anticipated having to move their best stuff upstairs at some point, get their boats out on Main Street for a few days, and then, when the water went down, spend time cleaning up all the mud and damage at the ground-floor levels. When flooding got too bad, people were accustomed to seeking higher ground to wait it out. And there was plenty of high ground.

So the parade formed as usual in the center of town. The veterans in their Civil War uniforms, with the crowd of other citizens, began moving, the spectators carrying flowers, on the road along the Stony Creek toward Sandyvale Cemetery.

Up at the South Fork Club, meanwhile, there were few people around. The resort was largely empty, a summer retreat still in winter hibernation. Preseason maintenance was going on, and while the summer season wouldn't get going till mid-June, a few of the members had taken Decoration Day as an excuse for a four-day weekend and had ventured up to the club for an early glimpse of summer.

It wasn't raining there yet, either. But it was blowing. Under dark, low-flying clouds the lake's grayness was whipping up in whitecaps, even waves. Big trees on the hillsides above were swaying wildly, their leaves shaking. The few members on the scene watched as the wind swept across the big lake. They were in their cottages or in front of the big, crackling clubhouse hearth, protected from the winds that raged about the grounds.

Huddled against the wind in the blowing woods was a camp of recent Italian immigrants, hired to put in a new improvement: indoor plumbing. That project was being overseen by one John Parke, staying in the clubhouse at this nearly deserted resort. Parke was young, only twenty-two, but having studied civil engineering at the University of Pennsylvania, he'd been working for an engineering and architecture firm in Pittsburgh. Under that firm's auspices he'd been embedded with the big South Fork client to oversee putting in the new indoor-plumbing system.

William Boyer was here, too. His title was superintendent of property and grounds, and Boyer lived at the club, with the assignment to keep an eye on the property, serve as a firm presence to deter any thieves and poachers, and make sure any club members arriving in the off-season got properly fed and taken care of.

The few members who had come up included the Dewitt Clinton Bidwells. Mr. Bidwell, sixty-one, a seller of dynamite and other explosives necessary to the coal-mining and railroad industry, was connected by the marriage of his daughter to the wealthy Childs family, and thus to Henry Frick; Bidwell also served as a director of the Marine National Bank of Pittsburgh. He and his wife had come up to the club for the holiday and were planning to leave the next day, Friday.

And Elias Unger was here for the preseason weekend. Unger had been in the deputation that had met Daniel Morrell's skeptical engineer, John Fulton, during the reconstruction of the dam and shown him around. Now, with the passing of Benjamin Ruff, Unger was serving as the club's president. Known as Colonel Unger, despite never having achieved any military rank, he came, like Andrew Carnegie, out of the Pennsylvania Railroad. Unger had also managed a summer resort at Cresson, and Pittsburgh's Union Depot Hotel Station, but he'd retired from all other jobs to dedicate himself to the hotel in Pittsburgh he owned with Carnegie. By this spring of 1889, most people saw him as largely retired, but Unger also served on the board of directors of the Woodruff Sleeping & Parlor Coach Company.

As president of the South Fork Club, Unger brought his long experience in hotel, restaurant, and resort management to making the club run properly for the enjoyment of an august clientele he'd both served and been a well-regarded member of for many years. He lived up here now, not in the clubhouse but in his own big farmhouse, perched on the hillside across the lake from the clubhouse, on the spillway side, near the dam. His porch had a commanding view of the lake and the distant hillsides across the lake. His cows grazed nearby.

Unger had arrived at the club on Thursday evening, having visited friends in Harrisburg, and resumed his job as boss on the scene. He had a lot to do. Along with overseeing Parke's management of the Italian workers putting in the indoor-plumbing system, there were many necessary preseason repairs to get done in time to allow up to two hundred guests to start arriving in a few weeks: nailing down loose boards on the boardwalks; clearing branches that had fallen during the late snows; repainting some of the cottages; cleaning and repairing the fleet of boats. Trains running from Pittsburgh to the nearby little town of South Fork brought supplies to be delivered by horse to the club. Under his expert management, the work had been going pretty smoothly.

None of the practiced managerial men on the site—Unger, Boyer, and young Parke—thought the weather was worthy of note, aside from wild wind. When Parke, riding his horse along the roadway on the top of the dam, observed the level of the lake, everything seemed fine: the water was more than seven feet below the top, more or less normal for an especially wet spring. Unger, for his part, thought the water level was actually surprisingly low.

At 4:00 in the afternoon, as the Decoration Day parade began returning to Johnstown, having left the veterans' graves gaily decorated, rain did start falling, gentle and fine. Then it stopped.

With the skies clearing for the evening, maybe there would be no flood after all, people began thinking. Maybe no rain, even.

Up on the mountain, young John Parke ate an evening meal in the clubhouse. He knew it had rained only when he came out of the clubhouse to have a last look around for the night. It was dark now, too dark to see much of anything, and it was when he stepped off the clubhouse porch and onto the boardwalk leading to the cottages that he noticed dampness.

The wind was still blowing hard. But the rain had come and gone and the night sky looked, if anything, less overcast than before. No sign of a storm coming. John Parke made his way in the darkness back to the clubhouse, went upstairs to his room, and fell asleep.

Getting up on Friday morning, people in Johnstown began admitting they'd never seen anything like it. And yet they started doing what they always did: coping.

The rain had started up again, around 9:00 P.M. on Thursday, and while at first it had only rained as gently as that brief shower of the afternoon, soon it had begun pouring with a gathering, and then astonishing, force amid a high, raging wind.

Some people slept through it, amazing others who had awakened to hear the storm and found it impossible to ignore. Rain fell astonishingly hard all night; then, at around 5:00 A.M., a landslide caved in a stable at a brewery, and that woke up even some who had been sleeping through the rain.

The rivers were roaring. Everybody who heard them, in that predawn darkness, knew what that sound meant.

The gray, rainy light of morning confirmed all worries. The Stony Creek and the Little Conemaugh raged and sloshed by, murky and brown, on each side of town, carrying heavy logs, windblown branches, and other debris with incredible speed toward their confluence at the Conemaugh proper. In none of the previous floods—even in the worst, only two years earlier—had both rivers risen like this at the same time.

Johnstown knew what to do. School was canceled. Even some of the mills shut down; others operated on short shifts and reduced crews. The idea was for everyone to get home and see to things. While this was worse than usual, they'd been here before.

Store owners were trying to stow inventory upstairs or otherwise protect it. As basements began filling, families turned to packing away foodstuffs. Both the Stony Creek and the Little Conemaugh were rising more than a foot per hour. Nobody had seen that before.

As streets began to turn more quickly than anyone could believe into deep creeks, the kids, as always, started splashing around in them, delighted and amazed. That's when little Gertrude Quinn's big brother Vincent began wading into the road, against their father's orders, to offer help to anyone who needed it. Gertrude started reaching out for the little ducklings swimming in her yard, the girl so wet now that Libby dragged her in and got her dried off and changed before James Quinn came home for dinner.

But it wasn't just the kids. A man named Charlie Dick gave his watch and cash to a neighbor and began riding logs down the stream for fun. Other grown men were doing the same.

But Charlie soon became alarmed. The water was rising too fast. He went home, and he told his wife it was time to get out. She thought he was crazy, but he insisted. They took their children to a friend's house, on Napoleon Street, on the north end of town, higher ground than the lower end. Charlie had gotten soaked fooling around in the water. Chilled, he undressed and went to bed and fell asleep.

Throughout the morning of the thirty-first, however, most people stayed pretty calm as they went about their work. This was the valley ethos, the sense that as high as both rivers were running today, flooding had happened before and would happen again and there was nothing to do but get through it the best they could.

Up on the mountain, in the little town of South Fork, so near the South Fork Club and the lake, people had spent Thursday night even more keenly aware of the storm than those it had awakened down in Johnstown. Sitting up in the dark, hoping to wait it out, some people on the mountain had the sense of being caught in a cloudburst like no other, with no beginning, middle, or end. A Mrs. Heidenfelter, resident of South Fork, called it "a rumbling, roaring sound . . . I thought the day of judgment had come when the roaring and the awful rain happened," she said. "People say the noise we heard was a waterspout, but I've never seen one and don't know how they act." Some reported thunderlike sounds that could not, they insisted, have been mere thunder. At a sawmill on South Fork Creek, logs that had lain in place for forty years were carried away by the swelling water. Arising having barely slept, people in that mountain town heard a roar from the streams and rivers that seemed relentless, and when they looked out on their world, what they saw was hard to believe. The fields were under three feet of water, and the streets in South Fork were high, muddy, fast-flowing streams.

In his big farmhouse above the South Fork Club's lake, Colonel Unger awoke at 6:00 A.M. and found himself amazed as well. As he looked out from his high farm, everything seemed to be underwater. That would have been an astonishing sight anyway, but given the lack of rain the day before, and what had seemed to him low water in the lake, for a long moment Unger simply couldn't understand what he was seeing. This was a flood like no other.

The dam. Unger pulled on gumboots and threw on a rain-coat and splashed quickly down through his fields to that mighty earthwork. What he saw there was, again, hard to understand at first, even difficult to see. Overnight the water had risen until it was now maybe only five feet below the top of the dam, and it was still rising, visibly, at an amazing rate. As Unger stood there watching and calculating, the rise seemed to him to be happening at about ten inches an hour.

At this rate, the dam must soon be topped. While Unger was no engineer, he knew what that meant: water passing steadily over the top of an earthen dam and falling down the outer face will quickly wash away the stone coating, admit water into the earthen core itself, and weaken the entire structure. Sooner or later, the dam will then subside, releasing, all at once, the water it's been holding back.

If this flood now kept rising, and the dam were topped, the dam wouldn't just leak, or breach, but would actually fail. The South Fork Club would lose its lake. With the lake would go the prize stock of black bass. That was Unger's first concern.

It was critical, Unger began to see, to somehow release water from the rapidly rising lake and keep it from topping the dam. The time was now about 8:00 A.M. The water was only four feet below the top, still rising. The spillway was clearly overwhelmed.

But Unger was developing a plan. He sent for John Parke.

Parke had meanwhile arisen, too, at about 6:30 A.M. The youth having slept through a more pounding rain than anyone up on the mountainside could recall, he now saw from his window in the clubhouse that heavy rain had fallen. He went hurriedly downstairs and out onto the porch. What struck Parke as he looked out at the lake was of course the sudden, overnight rise of the water, but even more powerfully, a sound: a roaring like that of a waterfall the size of Niagara.

The roar must be coming from the inrush of the high South Fork, and the other five rivers, pouring an astonishing volume of water into the lake, at incredible speed. Parke went back inside and ate some breakfast. Then he found a rowboat and enlisted a worker from the plumbing crew to work the oars. They headed out onto the rising lake.

Parke wanted to inspect the South Fork and other creeks at the points where they were entering the lake beyond their normal bounds. When they got out on the water, the first thing he observed from the boat was that about a quarter of the lake, the segment nearest its intake from the South Fork Creek, had been filled with debris carried down and into the lake by the roaring rivers: big logs, sawmill slabs, planks. The force of the streams was pushing this huge junk pile into an eddy and holding it there, swirling wildly around the growing pile as water kept flowing in and the lake kept rising.

Then Parke was further startled. Looking down at the water's surface, he saw he was floating over the top of a fence. He knew this fence, four strands of barbed wire running, on a normal day, along the edge of cow pasture, well back from the shore of the lake. So the whole pasture was now well under the lake, which had spread far beyond its shoreline; the boat was passing over the fence with ease. Rowing past the smaller creeks' normal points of entry, Parke and the worker approached the main input to the lake at the South Fork Creek.

Here was a torrent. The South Fork wasn't a river anymore

but just a broad, rushing flood, rolling straight down through sloping woods whose trees stood in water many feet deep. This flood wasn't following any riverbed; it was just coming. Parke and the worker rowed to what had become the new edge of the lake, got out of the boat, and pulled it onto somewhat dry land. Then they started hiking upstream and uphill into the woods alongside the rushing flood.

They climbed about half a mile. They saw that the water was boiling, misting, pounding. It wasn't about to stop. They turned back down for the lake.

But their boat had moved, or rather, the shoreline had: the boat was already nearly adrift; that's how fast the water was rising and the lake spreading. They jumped in and went hard and straight across the water for the clubhouse.

Parke, trying to take into account everything he'd just seen, figured the water must be rising about an inch every ten minutes. On arriving at the clubhouse, he wasn't surprised to hear that Colonel Unger was down at the dam on the other side and had sent for him urgently. The young man ran for his horse in the stable, saddled up, and rode across the dam as fast as he could.

At about 8:30 A.M., William Boyer, the superintendent of grounds, drove by Unger in a spring wagon, taking Mr. and Mrs. Bidwell, along with a couple of other members, to the train station in the town of South Fork. Despite the flooding, Mr. Bidwell somehow hoped to get back to Pittsburgh today.

Leaving for South Fork station, Boyer asked Unger how things at the dam looked to Unger.

"Serious," Unger said.

Or did he? According to Mr. Bidwell's account, his reply was "Not serious." Later, that difference would matter greatly.

RAIN

When Boyer and the Bidwells and the other passengers arrived at the depot in South Fork, on a street that had already turned into a river, the local agent for the Pennsylvania Railroad asked them what was going on at the dam. Boyer assured the agent that while the lake was indeed rising, there was no danger of the water topping the dam. Mr. Bidwell pointed out, for the benefit of the agent and others gathering about, that for the dam to break, the water would have to rise four feet more than it had when they'd left it. That, the club member asserted, was an impossibility. The railroad man and the others at the station took this as good news from an authoritative source.

It was obvious from the flooding in South Fork, however, that the Bidwells' train wasn't going to be leaving for Pittsburgh anytime soon. Boyer left the industrialist and his wife and friends at the depot and headed back for the club. A number of men and boys of South Fork, perhaps not fully convinced by the good news, or maybe just curious, headed for the club as well. They wanted to get a look.

By the time John Parke had ridden, as fast as he could, to the dam and across its road to the spillway where Colonel Unger was standing, about fifty people had gathered on the hillsides above the dam. Near the spillway stood a bunch of men and boys from the nearby town of South Fork. The crew of Italian laborers was on the scene, too, ready to carry out Unger's new plan. A few more club members who had also come up for the weekend stood there, watching with amazement. The water level was only about two feet below the top of the dam.

With Parke there on the scene reporting the astonishing force of entry at the head of the lake, Unger told him the plan. The spillway was overwhelmed, and the water was certain to top the dam. The only thing to do was to cut another exit and release water that way: maybe with another escape route, water could pour out faster than it was coming in. Unger had sent for digging and cutting tools and was putting the team to work on carving out of the hillside, on the other end, the western side, an impromptu new spillway.

And yet in the midst of this desperate operation, Unger continued to think like the manager of a resort. The original spillway was so badly overwhelmed in part because of the layers of weir, wire fish screen, and angled logs intended to block the exit of the bass. That fishguard was now keeping water from leaving as quickly as it otherwise might. Added to that obstruction, a mass of debris was also beginning to be swept up against the fishguard—mainly tree branches and leaves—further slowing the flow of water out of the rising lake.

As the crew started cutting into the earth on the other side of the dam to create a new spillway, a man from South Fork named John Buchanan approached Colonel Unger. Buchanan pointed out the blockage caused by the fishguard at the original spillway. Even the trestle bridge over the spillway was causing obstruction, Buchanan said: the iron rods of the fishguard were suspended from the bridge, holding debris and blocking exit.

Buchanan told Unger that both the bridge and the whole fishguard apparatus should be taken down—fast. The men digging the new spillway wouldn't be able to get any deeper than the clay level; they didn't have the tools for cutting into rock. The only real chance for defeating the inexorable rise of the lake was to remove all obstruction at the original spillway and hope it would release enough water to lower the lake and relieve the dam.

And yet removing the fishguard would let the bass out. They were the pride of the club, Unger's chief concern. Buchanan argued, but Unger stuck to his guns. He made a fateful decision. He wouldn't take down the fishguard.

Meanwhile the hard, fast, wet work of cutting the new spillway went on, with water rising inexorably from behind. A related effort was under way to raise the height of the dam: men with plows and shovels went out on the top to build up earth there. But the roar from the head of the lake three miles away wouldn't let up, and the lake kept rising, quickly, against the dam. It was approaching the top.

At about 11:00 A.M., it became obvious to Unger, Parke, and Boyer—who had returned from dropping Bidwell and the others at South Fork—that keeping the lake from cresting the dam would be impossible. Really, the water had already topped it: as water splashed upward onto the dam's roadway, all attempts to build up the top were washing away fast. The efforts to cut a new spillway were having little effect, too. People standing on the hillsides above the outer face of the dam could already see leakage coming through near the base. On the inside of the dam, the roar of the water coming in from the South Fork and the other rivers was unabated.

It was clear to all of them now. In all likelihood, the South Fork dam was about to break.

Unger was even starting to reconsider his position on removing the fishguards—but it was late in the game to think about that now. At this point, removal might not help. For the first time that day, Unger began to see beyond potential losses to the South Fork Club: the loss of its lake, the loss of its bass.

Now he thought about the people below. They had to be warned.

PART II

WHEN THE DAM BROKE

CHAPTER 6

TAP-TAP-TAP

THE PEOPLE BELOW THE DAM WERE BUSY COPING WITH THE worst flooding they'd ever known. This coping was going on not only in Johnstown but in all the towns down the steep course of the Little Conemaugh: the mountain town of South Fork itself, elevation 1,480 feet; then Mineral Point, elevation 1,375; East Conemaugh at 1,309; and down at the bottom, in the hole, Woodvale and Johnstown itself, at 1,142. Getting word to all of them that they were in danger from something worse than any flood, that 20 million tons of water might soon be rushing down the valley at them all at once, had become an urgent priority to the men up at the dam.

But this late effort to give warning was to meet with a string of complications. Some of those complications arose from what had gone on, that long morning, up at the dam; from Unger's refusal to face reality, his misguided effort to save the lake and the fish without giving any real thought to people. Some, however, arose from the way people in the valley had become accustomed to thinking about—and not thinking about—danger.

Colonel Unger's first move was to get the bad news to the nearby mountain town of South Fork. From there, he hoped, a warning could travel almost instantaneously down the valley by telegraph: to Johnstown, lying so exposed at the bottom, and also to Pittsburgh, where Robert Pitcairn, the railroad executive and one of the most important members of the club, would need to be alerted.

The club had a direct telephone line to South Fork. But it was still shut off for the winter. So Unger asked John Parke to ride over to town as fast as he could and give the warning and get the telegraphy going.

Parke lost no time in mounting up, and he urged his horse over a wet and slippery road almost washed out in places by the flooding. Arriving down at South Fork in only about ten minutes, at around 11:30 A.M., he was riding as fast as he could in the river that Railroad Street had become, approaching the train station and the telegraph tower. Seeing a small crowd of South Forkers gathered in the high water in front of a hardware store across from the station, Parke addressed the group urgently.

Water, he told them, was running across the top of the dam at the lake. There was very great danger of its giving way.

People listened politely. But they didn't know this man, and he seemed awfully young. Anyway, the club member Mr. Bidwell, having been dropped off by Boyer at the station earlier that day—Bidwell was there now, waiting for floods to subside and trains to run—had insisted that there was no danger at all.

Still Parke pressed his case with all the urgency he could, and he asked one or two men to go to the telegraph tower next to the station and have the operator there alert Johnstown, the other towns along the valley, and Mr. Pitcairn in Pittsburgh. Then the young man turned his horse back up the valley toward the club, to help as best he could. By now, he fully expected to be met on the way with a wall of water unleashed by the dam.

From the train station, Mr. Bidwell had been watching Parke's exhortations. It all seemed a bit much to the dynamite executive. As Parke rode off, some of the people came in to tell Bidwell there was word of danger from the lake. He reminded them jovially that there was really nothing to worry about.

One man did, however, go to the telegraph tower and climb up to the office. Towers were still in use as telegraph offices: once needed to relay manual semaphore signals and flashing-light codes among high points, the towers had long since been converted to the electrical wire system demonstrated by Samuel Morse fifty years earlier and since adopted as a standard. The telephone, in use for more than a decade, had by no means replaced the telegraph: phones had the advantage of being installed in homes, but they could be inefficient, requiring manual exchanges, with operators patching calls from one party to another. Telegraphy could be performed nearly instantaneously, and a message could go to multiple recipients. The telegraph office was one of the most important places in any town, and South Fork was no exception.

That noontime, Emma Ehrenfeld was working in the South Fork telegraph tower. All the long morning she'd been receiving orders to hold and reroute trains because of the flooding. Now a man she barely knew, with a name she'd always been unsure of—Wetzengreist or some such—was in her office excitedly spouting off about big danger from the dam. Emma knew only that this Wetzengreist wasn't somebody people around South Fork took seriously, nor was he giving her a message to send. He was just ranting about danger and the need to alert people down in Johnstown.

Edward Bennett, the engineer for the No. 1165 freight, was in

Emma's tower, sheltering from the rain and chatting. Exhausted railroad men stuck at South Fork had been coming in for some time now, their trains stopped by rain, trying to figure out what was what. There was a coal stove on the ground floor of the tower, and Bennett was taking cover there with his conductor, S. W. Keltz, and trying to get dry while this strange-seeming man raved on about the dam.

Bennett had gone through a hard night. He'd been taking a night train of iron ore from Bolivar to East Conemaugh, and when the rain started, he'd tried to back the train onto what they called the "horn" at the East Conemaugh station, when three of his four cars jumped the track. He waited, while the wreck force was called in, but it had started raining incredibly hard. The crew didn't reach him until around 2:30 in the morning. It couldn't even get his cars back on the track until between 5:00 and 6:00 A.M.

So Bennett pulled up on the siding to move out of the way of the No. 2 freight and a passenger train, then got off the siding and followed them up to the South Fork station, stopped there, and waited for orders. By then Emma had begun holding trains because of the amazing flooding. Seeing that there was a train already on the middle siding, Bennett backed his train in behind the office, left it there with his fireman and flagman on the locomotive, and went into the tower with Keltz the conductor to warm up and find some company.

The flooding had risen so high now that Bennett was telling Keltz that if the dispatchers would move the Limited up—it was sitting on the track in the way—he would try to save his engine by running it up the valley to higher ground.

But now the man Wetzengreist was raving at Emma about the dam at South Fork, and Emma was trying to explain that the telegraph poles west of the next tower, at Mineral Point, had already subsided into the flooding: she couldn't get anything past Mineral Point, couldn't send anything at all directly through to Johnstown, or to Pittsburgh.

Still, Emma found the man very insistent. She wasn't sure what to do.

So she tapped out a message to the more experienced operator at Mineral Point, asking him for advice. Telegraph operators had these tap-tap-tap conversations all the time, adeptly translating code into text; long-distance code romances had even sprung up between bored operators in far-flung outposts. The Mineral Point operator tapped Emma back a suggestion that they err on the side of caution. As the poles west of his office were in the water, he should write down a physical message on paper, saying something about news of danger at the dam, and send it by foot to the railroad station boss at East Conemaugh, who might send a man or two down to Johnstown.

Emma tapped back agreement. But she and the Mineral Point operator had no real information, no official message they were charged with sending; they didn't want to be alarmist. Together they (tap-tap-tap) whipped up and (tap-tap) mutually agreed to wording that relayed what Emma had heard. The Mineral Point operator wrote the message down on two pieces of paper, hoping to increase the odds of delivery by getting the note into the hands of two separate couriers.

But he couldn't leave his post. So now the Mineral Point operator had to wait and hope somebody would come along and agree to carry the message down the line.

Eventually someone did, and that message, amazingly enough, made it all the way down to Johnstown. It got there not only on foot, in the end, but also by wire. Handed by the Mineral Point operator to a soaking-wet trackman, coincidentally sent up from East Conemaugh to check a flooded-out landslide on the rails, the message was handed by that trackman to his own boss, who carried it in his pocket toward East Conemaugh, then stopped on the way at another telegraph tower. That office turned out to still have a working westward line. And so this message, having begun as tapped-out electrical impulses, then converted to jottings on paper, and now returned to electrical impulses, was tapped almost instantly to the telegraph office at the East Conemaugh rail yard, and from there not only to Mr. Pitcairn's office in Pittsburgh but also to the railroad telegraph office in what was fast becoming the underwater town of Johnstown, Pennsylvania.

The message from Emma arrived around 1:00 P.M. As received at last by the Johnstown telegraph office, it went something like this: "South Fork dam is liable to break: notify the people of Johnstown to prepare for the worst." And it was signed, as Emma and the Mineral Point operator had agreed, only "Operator."

Frank Deckert was the Johnstown station's telegraph boss. When informed by a subordinate that this message had come in, Frank didn't read it or give it much thought. The message wasn't exactly authoritative. It had no source, no sender. "Notify the people of Johnstown" was a vague and sweeping suggestion; "prepare for the worst" was the stuff people said about that dam every spring. When an assistant ticket agent showed the message to a couple of men hanging around the station, they broke up laughing.

Such was the response to the earliest warning to make its bumpy and circuitous way from those anxious men up at the dam to the busy people of Johnstown. First John Parke on horseback, then the excitable Wetzengreist, then Emma in her tap-tap-tap with Mineral Point, then the railroad men by rail and foot, then another telegraph operator, and finally the yardmaster at East Conemaugh: together, they all did manage to send word, and in only about an hour's time, all the way down to Frank Deckert in Johnstown. But Deckert wasn't the only person in town too busy preparing for what he and everybody else thought of as the worst flood they'd ever had to live through to give much attention to the news of the risk posed by the dam.

By now the town at the bottom of the valley was in a state of constant activity, and immediate tasks seemed far more real than worries about something that had never happened. People's basements had long since gone underwater. Many of the ground floors now had inches of water, with more coming, and in the lower part of town, things were even worse. The water was many feet high and rising.

This flood wasn't just water. It was full of mud and waste, and everybody with carpets was pulling them up; those with nice furniture were hoisting it upstairs. Farmers meanwhile were leading as much livestock as possible up and away from the rivers onto high ground. In the flood that had occurred in 1887, quickly turning now into the second-worst flood ever seen in Johnstown, the waterline had topped out at about a foot above the floorboards of the ground floors in the lower part of town. This one would obviously be higher, although nobody could say how high. At offices and in homes, people sought one way or another to get important papers, books, and beloved objects well above that line. The idea was just to hunker down in the dark— all electricity could be expected to fail—and wait out the flood.

Other residents, however, were leaving, mainly in concern about the novel intensity of this flooding. Hundreds of people had come out into the muddy, flowing streets, some wading on foot, others getting their horses to pull wagons with wheels partly underwater. People loaded up food and valuables in parcels and were carrying them above the rising waterline, some simply holding the stuff high over their heads, others with packages lashed to wagons, others guiding boats and improvised floats like boards along the water's surface. Children and pets were piling out of windows into rowboats.

Only some part of this exodus was toward higher ground. Other people were hoping to check into downtown hotels. Among those thousands who were staying home, some looked up from their preparations and rolled their eyes at these long lines of burdened, fearful refugee families.

In the office of the *Johnstown Tribune,* the editor George Swank had decided to live-report the worst flood ever seen in town. Swank didn't know when he'd get his next edition out, but whenever he did, he wanted it to include his minute-by-minute log of the wild events of May 31, 1889. He was an old-time newspaperman, having worked for the great editor Horace Greeley. He'd been wounded at the Battle of Gettysburg; he didn't scare easily.

Swank was watching out his office window, and at noon he reported that the streets up and back as far as Jackson were running with a yellow, devastating flood. He recorded what he'd learned from people telephoning him: the Poplar Street bridge had fallen. One man said that a cow had been standing on a dislodged pier of that bridge, and had then fallen into the Stony Creek. The bridge from Millville to Cambria City had collapsed into the Conemaugh, and at 1:30 a man called to say he was standing in water up to his middle on the first floor of his house.

While all this frantic work of preparation and exit of refugees went on, two more messages of warning from the men at the dam would get down to flooded Johnstown, later in the day. Finally some in town would start taking them more seriously.

But by then, many people were ensconced in their upper stories and attics, prepared to wait out the flooding in the dark. And by the time the last warning arrived, there was little that could have been done anyway.

When John Parke got back to the dam, after pleading with the citizens of South Fork to seek safety and get word of imminent disaster to other towns down the line, it was about noon. The dam hadn't broken. But the men at the dam were now watching a sheet of water slide right over the center of the broad top of a dam now mostly obscured by a surging lake looking to pour down the other side.

The new material thrown on top of the top was gone: it had done nothing. The new, ad hoc spillway was letting water out across twenty-five feet, but it was shallow and had no discernible effect on the steadily rising water level.

Colonel Unger had finally changed his mind about removing the fishguards from the original spillway. It was very late to make that decision. At least he did make it. Men were therefore up on the trestle bridge above the impeded waters at that exit, trying to rip out bridge floorboards. Down below, others were trying to get at the angled, nail-studded logs and the hanging iron with the wire mesh.

It was no use. So much debris choked the whole fishguard apparatus, pushing with such great force against the big drain, that the screen and iron and logs were all tangled and choked with tree limbs and even with entire small, uprooted trees. Given all that, and the force of the water itself, there was no way even to move the heavy guards.

And the workers were now refusing, sensibly enough, to go along the dam's top: with the inner face all basically underwater, the structure was manifestly insecure. Alone, John Parke walked out on the road on the dam's top. About 100 feet of the dam's 270-foot-long top was now six inches deep in water. He splashed along that wide, submerged surface to its very center. Here the water was deepest. That fact suggested to Parke that there had been a sag, invisible but significant, exactly above the place where the old culvert had once run and had then been filled in. He looked out over the valley below. He peered down at the dam's outer face. Water coming over the top and down the outer face had already carved gullies in that face. The rock coating wouldn't last long.

Parke wondered whether they should try cutting yet another spillway, this time right through the earthen dam itself, a deep wedge cut from the top over at one end, where things were more stable. The earth of the dam was far more cuttable than the rock of the hillside.

But such a cut would send a terrible amount of water downhill all at once. It would probably eventually collapse the whole dam. Still, the ploy might do somewhat less damage than if the whole dam collapsed at once.

No. That would be playing God—doing active, violent damage to life and property in hopes of doing, on balance, less than if the dam broke on its own. There was still at least a chance that the dam wouldn't break. . . .

By about 2:30, however, with a large crowd now gathered on the hillsides in the rain to watch the lake slide across the road at the top of the dam and cascade down the other side, Parke, Unger, and Boyer were just standing there, too. Watching and hoping was literally all anyone could do. Observing the outer face of the dam, they saw the big rocks begin to peel away into the rushing stream below and fall down the valley.

And they saw something else: the water going over the top of the dam and sliding down the outer face was cutting great slices of earth from the face, thus digging a kind of hole in the dam. Water was slicing a big step, about ten feet wide and four feet deep, out of the mighty earthwork, and with more water always coming, deeper slices kept getting cut off, and the big, carved-out step kept getting larger as everybody gathered there just kept watching. It was ten minutes to 3:00.

George Johnston, a lumber merchant from Pittsburgh, had come to Johnstown to place orders. At about 3:00 P.M., he'd just arrived in the flooded town and noticed a bulletin, posted in front of the telegraph office. A crowd had gathered around it. George pushed his way into the crowd and read the news: the dam above would soon give way.

He got it. He knew Johnstown. His life, George thought, wouldn't be worth a snap once that dam gave way. And yet the people reading the notice seemed unconcerned.

Not George. He canceled appointments and started splashing for the train station, seeing the upper stories of houses fill with people and wagons piled with furniture pushing through the water in the streets. Some were heading for the surrounding hillsides.

Editor Swank was still typing.

Two more messages of warning had come down the mountain, and Frank Deckert, in the telegraph office, had begun to realize the situation up there was serious. Frank picked up the telephone and called the Western Union office. Mrs. Ogle, the telegraph operator at Western Union, was sitting in a room whose waters were high. Hetty Ogle felt she was needed, however, and so she stayed, tapping out the news she received of washouts and train reschedulings. When she got word from Frank Deckert that the dam was in danger, she relayed the message to editor Swank at the *Tribune*.

By 3:00 P.M., Unger, Boyer, Parke, and all the workers and local bystanders were standing on the hillsides and staring, transfixed by anticipation, at the dam. They could tell what was about to happen. It wasn't the amazing weight of the high water rushing into the lake behind the dam that would break it, as if pushing so hard that the dam couldn't withstand the pressure. It was the water that, having now fully topped the dam, was running down its outer face. That water was knocking off the rocky outer coating on that face and cutting off the earthen core in great slices, opening a widening hole, letting water out in increasing volume. The cutting would make the whole dam give way.

And as the people on the hillsides watched, the dam didn't seem to break. Having been cut open, the dam at last seemed simply to leave, to step away, to lower, all at once.

After the long, gradual slicing away of the core, there was a sudden, awful sound, rising quickly to a deafening roar. The dam was gone. That sound was the rush of the entire long, wide, deep lake heading out and down into the valley.

To Colonel Unger, that sound and sight were like a visitation of the vengeful, almighty God Himself. He fell to the ground.

At 3:15, editor Swank, having heard from Hetty Ogle at Western Union that the dam up the mountain was in danger, noted the information in his ongoing live-reported log. The editor considered, while pecking on his typewriter, the impossibility of imagining what might happen to Johnstown if the lake up at the club came down the valley.

He didn't know, as he typed that strange thought, that a towering, misting wall of water had been unleashed about five minutes before, didn't know the tower of water was roaring toward him, and toward all the people in Johnstown, on a twisting, fourteen-mile course.

Over at Western Union, Mrs. Ogle was pretty much done for the day. The water had now risen to the table where her keypad and machine sat: the wires were about to be grounded, the system soon to fail. There was nothing else she could do, so she tapped these words to the stations down the line to the west:

"This is my last message."

She meant her final communication of the day. She went upstairs to wait for the floodwaters to stop rising and start to subside.

CHAPTER 7

A MONSTER
UNCHAINED

Pₐᵣₖₑ ᴀɴᴅ ᴛʜᴇ ᴏᴛʜᴇʀs sᴛᴏᴏᴅ ᴛʜᴇʀᴇ ʜᴇʟᴘʟᴇss, ᴏɴ ᴛʜᴇ ʙᴀɴᴋs of the former lake, with Colonel Unger overcome by emotion on the wet ground, watching in horrified amazement as the event went on, mercilessly sudden, pitilessly long. The water rushed, as tons of pent-up water will, when so abruptly released from a great height, as if it had conscious will, fury, desire. With a force beyond anything they could have imagined before seeing it, the lake was emptying into the valley at an incredible rate and an amazing decibel level, tumbling over itself to get down there. Nobody could do a thing about it now.

They looked out over the valley and saw what the great, mucky, roaring, misting, rolling wall of water was doing with astonishing speed. Hundreds of great trees below them were ripped from the waterlogged ground, some uprooted instantaneously, others cracked off just as quickly, the whole forest on both sides of the monster pulled into the current and downhill with it.

But the land wasn't just being stripped of trees. The water was scouring the terrain of grass, turf, and soil. It was leaving only muddy, bare rock.

Just below the dam site was a farmhouse, and those observing from the hillsides watched that house, too, disappear before you could blink, its fallen boards added to a huge cauldron of moving debris that the flood was already picking up: fencing, vegetation, lumber, rocks, all swept into the torrent as it moved down the valley. Next they saw the flood take down Lamb's Bridge, just as easily as it had cut down the forest and the farmhouse. With the broken bridge added to the mighty tonnage of debris, the water turned with the valley. It disappeared from the view of the men at the site of the former dam.

The monster was on an accelerating rampage now, heading for the towns below. Having seen what they'd seen, they could only imagine the violence of the destruction about to occur. Below them, where the water had traveled, were no trees, no farmhouse, no bridge, the land itself stripped of all vegetation and soil, just Allegheny rock slicked by muck. The whole emptying process had gone on for a little more than half an hour. That's all it took—some said more like forty-five minutes—for the whole lake to be gone and out of view. During the exit, there was no way for those gathered by the original spillway with the prostrate Unger to get back to the clubhouse. They couldn't cross the racing water, and the road to South Fork was underwater now, too. So they stood where they were until it was over.

With that towering, misting wall gone elsewhere now and still moving, they knew, still raging and destroying, all was strangely quiet at the former dam site. The rain still poured, but nothing but acres of mud stretched far and wide below the hills where minutes before the lake had been. A stream wandered on the bottom, far below the hillsides, at high speed, through a vast, low expanse of muck. That was the South Fork Creek.

Prize sport fish were flopping around down in that mud. As a gentleman's pastime, black-bass fishing was pretty much over for good here. Members of the work crew, fishermen not for pastime but for survival, climbed down into the mud and started gathering up those jumping, struggling fish.

John Parke and others lifted the dazed Colonel Unger. He was still in shock, not functioning at all. They began helping him toward the clubhouse and to bed. Parke, having tried to warn the people of South Fork, believed he might have been successful. But he knew there was no way of telling what might have already happened to that town by now, or whether word had ever gone from South Fork down the valley.

A little after 3:00 P.M., the freight engineer Bennett was still up in Emma's telegraph tower by the railroad station in South Fork, along with conductor Keltz, when they saw people running like mad. There was no time to gawk in astonishment: from the tower's height the two men got an ideal view of something they'd never seen before, and couldn't understand, as it rushed down from the heights and entered the town.

Bennett and Keltz were yelling to Emma. They started charging down the stairs. Emma looked out the window and saw something like a moving mountain, coming straight for the tower and getting close.

People below were running. Some yelled up at her to move.

Without even stopping to get her hat—that's how scared she was—Emma ran down the stairs. She flew across the tracks to the steps of a coal tipple—a building storing coal, with a chute for sending coal down to an engine. She ran up those steps and turned to look back across the tracks at her telegraph tower. She watched the moving mountain sweep the tower clean away.

Bennett had meanwhile begun carrying out his plan to save his locomotive. He'd said earlier that if they moved the Limited up, he would do so, and now the locomotive was in even greater danger—the mountain of water would take it away—and as Bennett started for it, he could see his flagman and fireman lying on the engine, asleep. They had to move the locomotive, and move themselves, or be drowned.

Bennett ran to the engine and jumped on. The other two, not only roused from sleep but instantly frantic, joined his desperate effort. Though low on steam, they fired up the engine, started it chugging, and pulled off the siding and began running the engine and its four loaded freight cars across the bridge, the flood roaring by on their left. If they could make the other side, they'd be farther from the moving wall of destruction, and they did make it, but just as they reached the far side of the bridge a gigantic tree, shot from the nearby flood, banged into the engine, knocking it partly off the track.

Bennett closed his eyes. It was all over, he figured. The engine, still moving forward but partly derailed, was about to go off the bridge and into the flood.

He opened his eyes. A local freight train was pulling out of a siding about twenty-five yards ahead of him onto the track, as Bennett's engine kept limping forward. Bennett realized he was about to run right into that freight. He threw his engine into reverse, to get it to slow and avoid hitting the local freight, and when he came at last to a stop he yelled to his crew. They all jumped off the engine onto the tracks.

On foot now, the crew raced for high ground in South Fork, while Bennett started running up the railroad tracks toward the local freight still going slowly up the tracks on the other side of the bridge, also headed for high ground. Wading in three feet of water as debris surged past him, Bennett caught up to the moving freight train. He reached up for the second car from the caboose and hoisted himself aboard. He looked back at his own train.

When the wave hit his locomotive, it survived. But the whole train behind it was suddenly swept away, all four cars, the two of coke and the two of baled hay, before Bennett's eyes. There was nothing left but the locomotive.

Yet in South Fork, that moving mountain of water had soon come and gone and done little damage, compared to what it was about to do elsewhere. Only as the water fell past that town, and tumbled downward and picked up speed and momentum, and more and more debris, did it begin to wreak havoc that would rarely be matched in the annals of American disasters.

Tracking the former lake's movement from its dam site all the way down to Johnstown is like tracking the course of a crazed, gigantic, unimaginably powerful force. The thing seemed more like a visitation from a hostile otherworld than like a flood, a monster plowing a path, bent only on destruction, reducing everything in it way to nonexistence.

Its course followed a way carved over eons: the valley of the Little Conemaugh, beginning at the town of South Fork, where the South Fork Creek, so recently dammed, flowed into the Little Conemaugh. It continued along that course on a drop of nearly five hundred feet over about fourteen twisty miles. Along the Little Conemaugh had been built those towns, at points successively lower: Mineral Point, East Conemaugh, Woodvale, and finally Johnstown, laid out flat on the bottom, open to the worst hit. The gigantic, racing monstrosity naturally followed that course.

But just as naturally, the thing wasn't contained by the banks of that course. Had the dam never broken, the day would have seemed epic anyway: a river already flooded beyond its banks to the point of stopping trains and destroying bridges had risen to fill the valley at a level higher than anybody had ever seen before. Such a river, flooded well beyond its own boundaries, could serve only as a general guide for the course of the huge, raging, new thing that now came crashing down on top of it.

As it accelerated down the valley, the thing was probably about forty feet wide at South Fork town, but its width was varying all the time it moved, determined by the varying widths of the valley itself, which also, by narrowing and widening, affected the height of the moving wall: the phenomenon swelled and flattened and rose and fell crazily as it came. It had begun by bearing great piles of debris along with it—at first the earthen architecture of the former dam itself—and as it came, scouring the land of everything growing or built in its way, that tonnage of debris grew, too, gathering even greater deadly force to drop on Johnstown at the bottom.

One of the most telling moments came early. A massive stone viaduct built by the railroad company stood high across the Little Conemaugh, just downstream from a big, abrupt bend in the river, an especially narrow stretch at normal times. Made of sandstone, the viaduct had been built as an element in the amazing portage segment of the old canal-rail system; in recent years it had been serving as a bridge for all trains passing through the town of South Fork. Its one, high, impressive arch at once let water pass beneath and connected the two high hillsides.

Now as that viaduct, in all its peaceful solidity, stood bridging the flooded river in the rain, there was a distant roar, louder and louder. Then, from around the abrupt turn just upstream, came the giant wall of water and debris, having hurtled out of South Fork, throwing itself toward the viaduct. By now the wave had passed out of a broader part of the valley and into a channel where the narrowing sides of the valley forced it to rise to about seventy-five feet as it raced. Coming around the tight bend about as high as the viaduct bridge, the towering thing hit the great stone structure with titanic force, pinning against the viaduct's arch tons of ripped-up steel rail and huge trees and everything else it was carrying, blocking the arch, making the viaduct yet another dam.

For entire minutes, the monster paused furiously in its progress, booming and churning as more of the lake water rushed in behind the improvised dam of debris and masonry, rising to nearly eighty foaming feet, fighting and straining to move forward. The struggle went on. An irresistible force had met an object that, it turned out after a long pause, was not immovable. The great structure collapsed entirely. After all that resistance, the viaduct took only seconds to go. It jumped into the debris that had pushed it down, as the tower of water, suddenly set free, sprang forward yet again.

During those long minutes of pause and strain, the water of the former lake had been given time to collect itself, bring up its reserves from the rear, and pile up as a single, forceful entity. To whatever degree its power might have been slightly lessened by the fall through the valley, which did force the wall to spread out slightly and to throw off some of its energy in sideways destruction, now the thing had regrouped, coiled like a spring, and launched through the viaduct site like a second dam break, this time even more explosively, as if launching out of a new starting gate. With massive chunks of viaduct now sweeping down the valley, too, this rising, churning, plummeting, misting, roaring, battering force rushed on.

◆ ◆ ◆

A sawmill and a furniture factory, some houses for the inhabitants, and a railroad station: that was Mineral Point, a small mill town of about two hundred people, with houses lining the riverbank. Then the settlement vanished.

One minute Mineral Point was standing there in the rain, its riverbank flooded unusually deeply. In a few minutes of churning, crashing, bewildering destruction there was no more Mineral Point. Because nothing here was big enough even to pause the water, change it, or resist in any way, the town simply dissolved in the time it took for people to take a few adrenalized breaths. Every railroad locomotive sitting on the tracks, every length of those tracks themselves, every one of the thirty or so houses that lined the riverbank, every backyard garden, the sawmill, the factory: it all washed down the valley as the moving mountain, having shown up out of nowhere, moved on just as quickly.

Yet only sixteen people died in Mineral Point that day. The sheer extent and height of the flooding had already scared a lot of people out. With their ground floors underwater and their outbuildings starting to float and bob, most of the citizens had gotten out of town and up on the sides of the mountains. That's where they stood now, stunned, watching the wave crash down from above and wipe out their homes in a long instant before their eyes. And the water roared on.

Some of the Mineral Point residents had stayed in town. A man named Abraham Byers had wisely moved his wife and five of their six children out of their house onto high ground; his eldest, a son, got out on his own. But Byers's mother-in-law had refused to leave, and his wife, terrified for her mother's safety, returned to the house to plead with her mother to join the rest of the family. That's when the wave hit, and both women were killed.

A man named Christopher Gromley, in his attic with his family only moments before, suddenly found himself riding his rooftop with only one of his sons: their rooftop-boat was randomly surfing the huge, moving, rolling wall of water. They went speeding and bouncing down the churning water until they saw a place to jump off to safety. Back in the attic in Mineral Point, Gromley's wife and other children had been killed. The capricious monster, meanwhile, had moved on, down the valley.

In East Conemaugh, the next town on the flood's path, some of the citizens did get warning.

By now the wave was making an amazing, even a unique sound. It was a roar, of course, a rush of water amplified, but out of that noise emerged deep groanings and grindings, as water and debris beat up and tore at and removed entire landscapes.

John Hess, a railroad engineer running a work train out of the East Conemaugh yards, had been supervising a crew trying to take care of landslides on dangerous track nearly washed out by the flooding. They'd been working up near the telegraph tower at Mineral Point, to which Emma's first message had come from South Fork, so they'd heard general warnings that the dam at the club was under strain. Now they were back down the valley, working on track only five hundred yards or so from East Conemaugh station, on the Mineral Point side.

That's when they heard the sound. They looked up. It came from up the valley. They saw nothing, but Hess knew in a split second what that sound must be: close behind him and out of sight, a deadly wall of water. They were in its path, and they were the only thing between it and the whole town of East Conemaugh.

Hess knew what to do—maybe more accurately, he didn't

know what else to do. The crew jumped aboard, the fireman hit the steam box, and Hess chugged with everything he had into East Conemaugh, pulling his whistle open to a steady scream. As they went, he tied the whistle mechanism down so it wouldn't stop and came blasting into town that way, ahead of the wave, screeching out what he hoped would be a warning. He thought he could stay ahead of the wave.

And Hess did manage to get his train and crew to safety, while doing all he could to give warning to as many people in East Conemaugh as possible. He never saw the wave behind him. He only heard it, and it was a sound he would never forget.

Nevertheless, as the gigantic wave came roiling into East Conemaugh it caught many people unaware and leveled the town, just as it had Mineral Point. Ahead of it, a wind now blew. Survivors thought the gust alone pushed houses off their foundations before the water even reached them. This was a weird thing: mist but no water preceded the gigantic thing. It really was a wall now, and aside from its foaming yellow crest, so high above, it didn't even seem wet. It appeared solid, and yet its front was a conglomeration of trees, rocks, buildings, timbers, freight cars, all actually seeming to squirm as they advanced.

On a siding at East Conemaugh the Day Express had been stopped since 10:00 A.M. because of the flooding, along with two other passenger trains and a freight. Many of the passengers were outside in the rain, pacing up and down the tracks; others were inside, reading and looking out the windows.

As the day wore on, they'd seen the river working its way over their track. Their train pulled back and moved over to another track. The rails it had been on fell into the river. The train moved again. The second track fell into the river. Now it was late afternoon, and bridges had fallen into the river, and still they waited. There was nothing else to do.

Then they heard a shrill, long whistle, coming from some engine nearby. They looked up.

There it was: the huge mass of trees and debris and water coming on, looming over them only about two or three hundred feet away. The passengers on the trains fled the cars. Other trains were between them and safety, and some tried to go over, others crawled under.

Only one man rejected those options and started running down the track ahead of the wave to get around the whole set of four-car trains. He made it to the other side, jumped off the embankment with the wave now on the far side of the trains, fell into a deep ditch, landed on a plank spanning the ditch, and bolted for town. Looking back quickly, he saw most of the people trying to scramble over or under the trains, as well as those falling into the ditch, getting swept up in the wave. He turned and ran into the streets, turned back to look again, and saw houses floating away behind him and the flood chasing him.

He ran again, the water closing in on him, houses toppling over. Now the torrent seemed to be trying to head him off.

But he'd reached the hillside and started climbing. He was beyond the flood at last. Again he turned to look.

His train and the other trains were broken, swirling, cars loose and plunging downstream, two men on top, others still inside. Then all the trains started down the river. He yelled, in anguish. They flowed about five hundred feet, stopped strangely, and the engine of the train he'd been on reared up and came down on the engine of another train. Other mighty engines, pulled from the East Conemaugh roundhouse, were rolling against these trains now, the whole tangle of locomotives and cars backing up against a mass of trees. He could only stand there and watch.

✦ ✦ ✦

Woodvale, the next town down the valley from East Conemaugh, was a different kind of place. Its fanciful, romantic name reflected the difference: the Cambria Iron Works had built Woodvale as a "model town" for its workers and managers, and the town was the pride of the company, a kind of faux New England village, with a woolen mill—owned by Cambria Iron, of course—and pretty, white-painted clapboard houses. And unlike in East Conemaugh, the people of Woodvale heard no blasting train whistle, nothing to warn them of what was coming.

The big, mixed mass of water, with all the gigantic things it had collected, high above the normal river channel, poured right down on top of Woodvale. Stores, churches, and homes went fast, and chunks of buildings smashed against each other, drowning the people still trapped in crumbling attics or crushing them in tons of debris.

Some saw the towering wave sweep into town, looked up at it, looked over at the hillside, and saw that their way up was blocked by a long freight train sitting on the track below the hill. Desperate, they ran for the hillside anyway, threw themselves down on the track, and started crawling under the cars in hopes of getting to the other side and climbing to safety. The wave smacked the train. It started rolling. The people underneath were crushed to death by the train.

The woolen mill, meanwhile, started coming undone in the rush of water. Men in the mill kept climbing and moving from room to room as pieces of the place just fell apart. Some were left, at the end, clinging to the one, tiny corner of the building that still stood. And so they lived.

Yet it took less than ten minutes for the water to kill one out of three of Woodvale's residents, more than three hundred people. From one stable alone, about ninety horses were swept whinnying and kicking into the waters to die, and of course the stable went too, along with all the rest of what had been Woodvale, the pretty New England–style houses, 255 buildings in total, a tanning factory, and all but the one corner of the mill.

The water moved on, carrying all that tonnage. Johnstown, connected to Woodvale by a streetcar, was next in its path.

The town of Woodvale had moved on, leaving only a broad mudflat where the town had stood. It wasn't just that the buildings were missing. It was abruptly hard to believe that any buildings had ever stood here, or any trees or shrubs, that any track had ever been laid. Everything was gone.

Almost everything. Eerily, one piece of iron railroad bridge, no longer anchored to anything on either side, somehow loomed above the mud. Its bleak and lonely form was silhouetted against the sky, spanning nothing but the expanse of muddy emptiness and dead bodies that the flood had left.

And just north of Woodvale, a tiny town on the hillside, New Austria, with only about thirty houses, was left entirely intact. This was a game of inches. Where the wave destroyed, it did so utterly. Those just out of its path were untouched, but left to witness, amazed and horrified, all the mind-bending damage such a mass of water could do.

Having added Woodvale to its body, the monster just kept going.

Woodvale and Johnstown, almost overlapping on the map, were literally connected by a streetcar track from Woodvale's main street into Johnstown. On what was effectively the border of the two towns stood the Gautier wire works, owned by Cambria Iron.

Thanks to the flooding, the factory wasn't in full swing, but a skeleton crew was working and the boilers were making steam when the wave hit. A huge plume of steam was released into the air. Then the whole factory itself rose, for the moment still intact, into the wave and rode off on the water, lending its white, scalding mist of steam to the white, cold mist of the floodwaters. As the factory rushed along in the flow, it started breaking up. As the men caught inside began drowning in the churning water and getting smashed to death in the roiling debris, the factory added miles of barbed wire to the deadly mix of elements that the beast had become.

So it was that when the whole gigantic thing that had begun as 20 million tons of water, timber, rock, and an earthen dam, having picked up and brought with it whole forests of trees and boulders, a stone viaduct, bent steel rail, entire locomotives and train cars, a few towns' worth of buildings and factories, dead animals and people, and now thousands of steel barbs strung along miles of wire, came roaring and misting and steaming and pounding and slashing into Johnstown, with all the force gained by a nearly five-hundred-foot fall, it had become an indescribable presence beyond any rational belief. If this had been only an amazing quantity of water, badly dammed and now unchained, that would have been enough for a disaster. But this water had carried down the valley's entire industrial enterprise. The visible investment of the millionaires was swept forward on the wall of water, all the brick and iron and fire and steel that had generated those millions by drawing labor from the people who lived where the water now raged. Everything that had so changed the valley of the Conemaugh, in such a short time, had been not only destroyed but also weaponized for total destruction. And so the thing came into Johnstown.

CHAPTER 8

CAULDRON

Gertrude's Papa had gone back to the store after the midday dinner, having left his instructions to Aunt Abbie and the nurse Libby that the children were to be made ready to run to the hillside should he determine it was necessary. Aunt Abbie thought it was silly.

Meanwhile Gertrude, dauntless, had gotten out again.

She was dangling her feet in the water from the porch and watching the ducklings swim in the yard. It was strange to watch the pansies in what had been the flower garden, their stalks invisible, their faces twisting and turning just above the waterline.

Suddenly Papa appeared. He was splashing toward her, smoking a cigar, and Gertrude jumped up in hopes of getting ahead of him. But he took hold of her, and she bounded through the door as fast as she could while Papa landed blows from his open hand on her bottom with each step.

Inside, he gave her a serious lecture on wet feet, general obedience, and the danger that might be facing the family with the rising of the water.

The flooding had gotten bad enough that James Quinn's plan now was to take the family to the hill as quickly as possible. Vincent wasn't home: he'd stopped at his uncle's clothing store, just a few blocks down the street from the Quinn home, helping to move things there to higher levels. James had to hope the family there would look out for Vincent: James had to get the girls up to high ground.

In preparation, Libby started changing Gertrude one more time into dry clothes. James waited impatiently as the family gathered. Finishing his cigar, he went to toss the ashes outside and opened the front door. He saw white mist and heard a roar that froze him to the bone.

He turned back to the room.

"Run for your lives!" James ordered in the strangest tone of command that little Gertrude had ever heard him use. "Follow me, straight to the hill!"

"The baby with the measles!" someone yelled. James hurled himself up the stairs and returned in seconds with the baby wrapped in blankets. Terse now, and so with even greater firmness, he turned to Aunt Abbie.

"Follow me," James said again. "Don't stop for anything. Run for your lives. If the town is spared we can return, of course, but we can't take the chance of staying here, so follow me straight to the hill."

There was terror in his voice, yet clarity of command in his orders. Nobody else in the house had yet seen what James had seen, and the family followed him in panic, Aunt Abbie carrying her sleeping baby Richard, Libby carrying Gertrude. Nobody was questioning James Quinn now.

Or so it seemed, at first. Aunt Abbie, moving toward the door, had her doubts about this whole operation.

◆ ◆ ◆

The floodwaters had begun to rise more slowly, or so the Reverend David Beale was thinking. Indeed, they seemed to be starting to go down at last.

This was about 4:00 P.M., and the pastor of Johnstown's Presbyterian church was at his desk in his study, immersed in working on his sermon for Sunday, when he was interrupted by his wife, who believed the carpet in the parlor should be taken up immediately, to preserve it from the flooding. She was calling on the reverend to help.

Beale saw no reason to take up the carpet, and two neighbors were in the house, visiting, a Mr. Lloyd and his sister; still, the pastor put down his work to comply with his wife's wishes. And as he went into the parlor, he heard the sound.

A railroad train, somehow heading straight for his house at full speed? . . . Impossible.

Yet something strange . . .

The place exploded.

"Upstairs!" Beale yelled, as the water rose fast in his house. He shoved his wife and children and the two neighbors, plus Guess the family dog, ahead of him up the staircase, grabbing up the family Bible as he followed them.

But he didn't have to run: he found himself pushed straight up the stairs by the rising water, and since it got to the second story faster than he did, now he was nearly swimming. As he struggled to gain the upper floor, already waist-deep in surging water, a hat rack from the first floor, also thrown by the water, slammed him hard from behind.

As Beale's family rushed ahead of him toward the attic, a man shot through a window on the water into the second story.

"Who are you?" Beale said. "Where are you from?"

The man gasped, "Woodvale." The man had been carried all that way. Now he and Beale waded as fast as they could for the attic.

There it emerged that the man—Beale recognized him now, from the woolen mill—had been riding a torn-off piece of roof and had been thrown through Beale's window when that make-shift craft had struck the house with force. Beale and the other grown-ups now expected everyone here in the attic to die. Beale opened the Bible and read Psalm 46: "Therefore will not we fear, though the earth be removed, and though the mountains be carried into the midst of the sea. . . ." He passed the Bible around as the others shouted out prayers, and Beale led them as best he could.

"Though the waters thereof roar and be troubled, though the mountains shake with the swelling thereof . . ."

Beale's twelve-year-old son recited from memory the words of Psalm 23: "Yea, though I walk through the valley of the shadow of death, I will fear no evil: for thou art with me; thy rod and thy staff they comfort me.

"Amen," the boy added.

They had good reason to expect to die. For some random reason, the Beale house, though very precariously damaged, and seemingly about to collapse, wasn't yet in pieces in the water. So the group huddled in the attic was able to witness something not out of Psalms but out of the book of Revelation. The house stood in the midst of a raging sea, thick with gigantic detritus, and hundreds of houses and parts of houses were rushing and circling past the attic window on the water, bobbing and break-ing up, along with barns, freight cars, passenger cars, locomo-tives, trees, and animals, all sweeping by.

And human beings. Many were dead as they flowed and spun past; others were dying, still others struggling to live. Beale recognized them: there was Mr. Benshoff the bookseller; Mrs. Fulton and her daughter; Charles Barnes; Mrs. Young . . . they all rushed past, some already detritus themselves, others still hanging on as best they could to objects. Two small children, nearly naked, clung to a piece of roof. Past the window they went. A group of young women on another swirling, surfing roof were hugging each other. Beale could see that every other house over a long distance was off its foundation, falling apart in the water.

Now Beale saw another family he knew, a Captain Hart and his wife, his sister, and two children, struggling on some wreckage that was drifting near his house. Hoping to help, Beale splashed and swam to the stairs, plunged into the high water on the second floor, guided the Hart family into the house through a window there, and brought them all up to the now-crowded attic.

Wreckage had meanwhile begun to move away from the house. But now the whole house began shaking. Soon it was rocking. Back and forth they went: the house, they realized, must be coming off its foundation. Suddenly they could at once see and feel the whole structure starting to sag in the middle. The Beale house too was finally breaking apart in the surging tides.

Beale knew he had to act. Captain Hart and the neighbor Mr. Lloyd were insisting that the danger in the attic had become greater than the danger in the water, and Beale had to agree. The adults, talking fast, came to the conclusion that they had to escape. The only hope seemed to be to get onto the roof of the house, then try to use the roof as a float when the house finally broke up all the way.

Yet there was no way everybody in this attic could climb out of the attic window, only six or eight feet above a raging, deadly sea of debris, and scale the wall onto the roof.

Then they saw a roof floating just below the window, not

rushing past but bobbing and eddying and staying near them. They decided to go out the attic window, climb down, and get themselves and the children situated on that float. Beale grabbed a length of a rope that was lying around in the attic. He and Captain Hart each took an end. Hart made the first exit. He was to test the sturdiness of the craft.

Holding his end of the rope, Hart climbed out the window and down onto the floating roof. Still holding the rope, balancing on the big float, he yelled back to Beale that it seemed pretty strong. With the rope connecting the attic to the float, Beale began helping each person out of the window and into the water. Hart helped them onto their makeshift craft.

The two youngest Beales were pleading with Beale to save Guess. Beale lowered the dog onto the roof, too. Once everybody had gained the float and were bobbing around on it, Beale followed them.

Now they were a batch of human beings and a dog, floating on a roof. They surveyed their bizarre and terrifying surroundings. The movement of the water had already calmed somewhat. Maybe the flood had begun moving on. There was so much debris in the water that a lot of it had come to rest: mountains of wreckage were everywhere, weird, huge structures of sheer brokenness sticking out of the high water, everything piled onto or leaning against something else. In this part of town, there was more wreckage than water now.

They just might, the Beales and their friends thought, be able to climb, walk, and scramble through this new and freakish landscape and find some kind of safety somewhere.

George Johnston, the Pittsburgh lumberman who had read the notice of warning in the station, had canceled his appointments, and had splashed as fast as he could to the train station in hopes of getting out of town fast, had finally gained the depot. He was going up the steps to the platform, where the Pittsburgh train stood waiting, when he heard the roar.

Like so many others, George thought for a moment that the sound was that of a long, heavy freight train somehow racing toward him, and because he was at the train station, his delusion lasted a heartbeat or two longer than others'. But the sound was too loud and awful even for that.

Bewildered, George boarded the waiting train, sat in a window seat, looked out the window, and saw something impossible to register.

From up the Conemaugh came a moving wall, a yellow wall. Its crest was white.

Now Johnson saw that the white crest was frothy. He leaped up, but the train began to pull out of the station.

In a state of crazed terror, George fell back into his seat. Maybe this would be the safest place. As the train moved along the track above the town, all too slowly approaching the arched, stone railroad bridge across the Conemaugh, George could see that what had been the bustling yards of the sprawling Cambria mill was already a high, splashing yellow sea, rising fast, with houses and barns riding on it.

Now an eruption. At deafening volume, one explosion from the sea devouring the Cambria Works, then another, then more. All that molten steel, swamped by all that cold water, was simply blowing up, and more explosions followed, joining the cacophony of the water's roar and the banging and grinding of debris. George swung his head around to look the other way, down the valley. Water was rushing through the streets of Johnstown. Near his moving train, men and horses were floundering in the rising flood. The housetops were getting crowded with people, close enough to George that he could see their faces, deadly white. They clung to each other as the waters rose toward them with amazing speed.

Through the incredible din, George could hear the train's bell rope ringing again and again. The conductor was frantically demanding engine speed. The train accelerated. Now it was shooting across the railroad bridge. The conductor didn't stop ringing the bell, and the train kept picking up speed while George sat transfixed: houses circled past in the water under the bridge, and the bridge might, George thought, collapse under the train at any moment. The train was going so fast that it seemed to him they were leaping over those crashing, yellow waves below. Then it seemed that the train had left the rails, had left the surface of the earth, was flying through the air. And still they raced.

Suddenly, in the windows: forest-green trees blurred by the speed of the train. They'd gained the hillside opposite Johnstown, and George's heart leaped. The train didn't slow down. The pace seemed insane. Wondering if the engineer had gone mad with horror, George took one long look back.

Speeding along on the hillside, he could see the valley for miles in every direction. It was just a huge, boiling cauldron, everything underwater, with roofs and mill standpipes protruding. In one part of the former city, the water seemed to be piling up especially high. The worst, George thought, might be yet to come. He sat back and let his mind go blank as his train raced toward Pittsburgh like a demon out of hell.

Down in Cambria Iron's company store, a clerk named Thomas Magee was standing by the open safe at 4:15 when the water struck the building with a crash and poured in from every door and window, on all sides, even from the floors above.

Magee reached into the safe. He grabbed a tin box holding more than $12,000 in greenbacks. He raced with the cash to the second floor, other clerks right behind him.

No good: they were up to their waists in water. They raced up to the third floor. Much inventory was stored here, including bread and meat. Magee and the other clerks found canvas and oilcloth and cut off big pieces of it. They started wrapping the food in the packaging. Magee still had the cash.

Victor Heiser was sixteen, with an ambition to become a watch-maker, and that afternoon while the flooding rivers had been rising through town, he'd made a dash for the stable, at his father's request, to unhitch the two good horses his father set so much store by: his father feared they'd be drowned in the rising water if they couldn't escape. Victor had fought his way through two feet of water to get to the stable, which stood on somewhat higher ground than the Heiser home, and the rain was coming down so hard that he was quickly drenched.

Victor Heiser, too, heard the roar before he saw anything. He'd just freed the horses and was heading back into the rain when his ears were stunned by a sound like nothing he'd ever heard before, along with earth-shaking bangings and grindings.

He froze, totally confused, and as he looked back at the house he saw his parents looking at him from an upper window of the house. His father was motioning to Victor, motioning frantically, to move up to the stable's red tin peaked roof.

Victor knew what to do. To make some roof repairs a few days before, he'd fashioned a quick way up, and in seconds he was on the tin roof's high ridge.

From there he saw the wall advancing with speed down his diagonal street, and now he did recognize it as water. To Victor, the thing was a dark mass. Within it seethed houses, freight cars, trees, animals.

It arrived.

Victor saw the thing crush his home like an eggshell.

The house, with his parents inside it, simply disappeared before his eyes. Victor wondered how long it would take him to reach the other world, too. On the stable roof he glanced at his watch: 4:20 P.M. on the dot.

Then the thing hit the stable.

. . . 4:20 P.M.: a Mrs. Stevenson and her two daughters were charging up the stairs in their new home on Market Street, the water behind and ahead of them. They climbed onto a bed. It fell to pieces. But there was no way to get out of this room now and up to the attic: water was filling the room, lifting them, sending them all quickly toward the ceiling, as they fought to keep their heads above water.

Then the water was pressing them against the ceiling. And it was still coming.

Mrs. Stevenson and her daughters ripped into and banged on the ceiling plaster. They succeeded in making a crack. They bashed out a hole, then pulled plaster away in chunks and into the water, making a bigger hole. Even as the tide held them against the ceiling, soon to crush them to death, they found the thin strips of wooden lath to which the plaster had clung, nailed to floor joists above.

They grabbed the lath. They wrenched it out of the joists and broke it up. They ripped a big cavity over their heads and went upward into it, between the joists. The water was still rising.

The attic floorboards must have been pretty lightweight, because Mrs. Stevenson and her girls broke into their own attic, from the room below, pursued by the water. And by the time they got up there, the whole house was off its foundation and riding the waters like a huge boat, on the verge of breaking up. . . .

. . . Miss Minnie Chambers was heading home through the flooded streets to her house on Main Street, after seeing a friend. She was lifted by the sudden avalanche of water, taken off her feet and carried ever upward on its surging flood. Her skirts rose around her and gave her support, at first, until they became saturated, and with her skirts now dragging her down, Minnie gave herself up for dead. She started to go under.

Here came an entire boxcar, floating and diving, not far away. Minnie swam with everything she had. The boxcar's doorway was open, bobbing above the surface. With desperate effort, she reached for the doorway and managed to clamber inside the car.

Now as she rode the churning sea, her car banged, with horrible, jarring crashes, into other huge objects, throwing Minnie around and seeming about to fall apart with every next bump. She caught wild glimpses through the doorway. The water was full of men, women, and children. Some swam desperately toward her car in hopes of getting inside.

But no, each person was ripped away and disappeared and Minnie Chambers went on her wild ride alone.

Aunt Abbie paused at the top of the steps.

Libby was beside her, holding Gertrude in the new set of dry clothes. James Quinn had gone down his terrace into the flood, carrying his swaddled baby daughter with the measles, certain he was leading his endangered family members, all except Vincent, on a splashing escape to the hillside.

With James's back now turned, Abbie was rethinking this plan.

"I don't like to put my feet in that dirty water," Abbie told Libby. "We may catch cold, get sick, and die."

Libby said, "I'll do whatever you say."

At this, Gertrude screamed and started to struggle in Libby's arms. She could see Papa walking in the water, going away, with her sister Rosemary on one side and her sister Helen on the other, each holding one of his elbows.

She couldn't imagine anyone, even Aunt Abbie, not doing exactly what Papa had said: his tone and face had truly scared the girl, and now she could see Rosemary—so little that the water was up to her chin—insisting on holding a pointless umbrella over her head; Papa was urging her to drop it. Other people were down there in the street that was now a river, trying to flee, obstructing one another's progress. Papa clearly thought Abbie and Libby and the children were with him, too.

Gertrude started kicking Libby, desperate to get away and go to Papa. But Libby held her tight. Why wear good shoes into deep water?

Aunt Abbie said, "Let's run to the third floor. This big house will never move."

The women turned, went back inside, and ran up the staircases to the third floor, Aunt Abbie carrying Richard and Libby carrying an enraged and struggling Gertrude. When they got to the children's amazing play space on the third floor, they opened a window and looked out.

Now both Aunt Abbie and Libby started moaning in sheer panic. Nothing could have scared Gertrude more than that.

Unless it was what she saw out the window. Everything that everybody else saw that afternoon little Gertrude saw, too, taking it all in for one overwhelming moment. The pandemonium, the screams and cries and people trying to run, the parents dragging children with heads bobbing up and down in the water, the eyes of animals and human beings alike starkly staring. She saw the humanity of Johnstown compressed in a single, panicked mass, a wedge doing everything it could to get up to the hill. And along with the huge rumbling roar, bells rang, factory whistles screamed, steam engine throttles opened to shriek.

Gertrude saw the moving mass coming down on them. To her it wasn't yellow but black, darkened by houses, trees, boulders, logs, and rafters, a towering, crazy avalanche.

She heard Aunt Abbie saying, "Libby, this is the end of the world. We will all die together," and the two women were sobbing and moaning and starting to pray.

Gertrude became frantic. They prayed while Gertrude jumped up and down. She shouted, "Papa! Papa! Papa! Papa!" over and over again.

James Quinn had pushed through the waters with Rosemary and Helen and the swaddled baby, and when they reached the hillside and got above the floodwaters, he looked back for Abbie and Libby and the other children. He didn't see them.

James searched one terrified face after another as they poured toward him up the hill. No Abbie, no Libby. No Gertrude.

Then he knew. Abbie had gone her own way. Either they were too far behind or still in the house.

James became desperate. From the crowd, he picked out a young man he knew by the name of Kurtz, strong and solid. He handed the baby and the two girls over to Kurtz and started back down the hill against the flow of the fleeing refugees, toward his house.

It wasn't far. He could see the house. Soon he was back in the deep water on the street.

Then something stopped him. He looked up. He saw the wall of water coming down. It was as high as his house, and it carried everything from trees to churches, from iron beams to logs, a pile turning over and over itself, driving forward into town.

He saw, in one glance, all that was about to happen.

He was too late to save Gertrude from this.

Vincent was down there, too.

The towering thing was coming right at James. Just for a moment, in his despair, he thought he might simply jump into the monster. He might sacrifice himself to its rage.

But James turned back to the hill. In despair, he ran for his living daughters. The monster came down. James slipped and almost got hit by the wave, but in the game of inches he got up and ran and climbed, and the water dashed against the hillside, barely missing him.

He stopped and turned to look back. He looked at his house. It was deep in the flood now.

He saw it start to tremble. He saw it start to shake. Then the house started rocking on its foundation.

Gertrude's brother Vincent had been helping save inventory at his uncle's store, just down the street from the Quinn home. When the bells started ringing and the factory whistles started blowing, Vincent, his uncle, his cousin, and everybody ran to the lobby, where the staircase would take them to the third floor.

But looking up the street, Vincent saw his father leaving the house with the baby in his arms and two of the little girls by his side. Vincent jammed his straw hat on his head. He started into the street, now a mighty river at flood stage. His uncle grabbed his coat and held him by the arm. They had a face-off.

His uncle wanted Vincent to stay.

Vincent yanked his arm away. "I must help save the little girls!" he shouted. And off he went into the flood.

It took him a long time to get anywhere. By the time he was only a few doors from his house, the water was almost up to his neck. This was at the intersection of Main and Bedford, where the confluence was making a powerful current, and he was trying to push through it when the wall of water arrived.

Vincent went under. The monster brought its piles of buildings and iron and barbed wire and cattle and trees right down on top of the boy.

As the monster bashed onward, a man looking out a third-floor window saw Vincent's straw hat rise to the surface. He saw the hat spin, float, and wander off.

Victor Heiser, perched on the stable roof ridge, had just seen his parents disappear before his eyes when his house was crushed like an eggshell. The boy had found time only to glance at his watch before the avalanche hit the big stable, ripped it from its foundations, and started rolling the whole stable in the water, like rolling a barrel.

Victor found himself crawling and racing and grabbing. As the stable toppled end over end, down the rushing water, he managed, with each turn, to keep getting on top.

Up loomed a neighbor's house. It was coming right at Victor, right in the path of the tumbling stable, both structures moving at high speed. The collision would throw Victor off the stable and straight into the water.

So as the house neared the stable, he watched, concentrating intently, waiting as long as he could, knees bent. The house came on and on, straight for him now, and in the instant the house hit the stable, the youth launched himself into the air.

He landed on the roof of the oncoming house. He'd minimized the impact to himself and stayed afloat.

But the walls holding up the new roof where he now sat began to collapse, and Victor was suddenly plunging. Trying to scramble upward as that roof became a falling slope, he saw yet another house pop suddenly out of the water beside him. As the house he was riding gave way, Victor reached for the new house, grabbed its eaves, and swung away from the house collapsing under him.

Now he was just hanging over the water as the house raced along. He was gripping the shingled eaves, and this new house pitched through the water as it ran as if actually trying to shake him off. As Victor dangled, his grip weakened, and the house turned and bucked. Victor felt his fingernails dig into shingles softened by water.

Sooner or later, he knew, he would let go.

Aunt Abbie, overwhelmed with terror, said, "Run and get into that cupboard," and she meant the big wardrobe with storage for the children's playrooms. She ushered Libby and Gertrude inside the wardrobe and pushed them far back into its depths. Abbie herself stood with one hand on the doorknob and peered out as if they were hiding from some huge force of evil stalking them.

Aunt Abbie said it again: "We will all die together. Lord have mercy on us." Libby was sobbing.

Gertrude broke loose, pushed the wardrobe door open, and jumped up and down, screaming, "Papa! Papa! Papa!" but they pulled her back into the cupboard. Libby held her tight. Gertrude didn't stop saying it.

The house shuddered, once. Then it started dancing. Then it shook. Gertrude peered through the crack in the wardrobe door and saw the toy pots and pans swinging. Then the house jerked, and the little girl could see that Abbie's eyes showed only the whites.

Now dust and plaster were dropping on them, lots of it. They were all coughing and choking.

Then the floorboards below Aunt Abbie's feet burst open. Yellow water came up fast. It rose over them.

Gertrude, in the water and in the dark, reached for Libby's hand but found none. Her mouth was full of dirt, sticks, foul water. She spat, kicked, splashed, grabbed. She threw her head back to get it above the water and to keep water out of her mouth. She gasped for breath. Every time she got some air, she said, "Papa!"

Victor Heiser, dangling with his last grip from the eaves of a racing, bucking, plunging, collapsing house, could do no more.

He tried to dig in his fingernails, yet his grasp was failing of its own accord. And so it happened.

He let go.

A long, gut-wrenching drop. But not into the water: Victor landed hard on his belly on something solid, and when he gathered his wits he saw that he was on a piece of his same old tin stable roof, moving fast in the water. Lying on it, face forward, he used whatever grip he had left to cling to its thin ridges, riding the surface of the flood on his strange raft, as the wave he was riding destroyed everything it met. He heard screams. He knew people were dying all around him.

He saw the fruit dealer, Mr. Musante, with his wife and two children riding on their old barn floor. As they raced along, they had a big trunk open beside them: they were packing it with possessions as fast as they could. Then wreckage heaved up, bore down on the Musantes, and crushed them all to death while Victor watched.

He looked ahead. He was being carried at top speed toward a big jam of wreckage: a pileup of huge debris against a stone church and a three-story brick building. Moving, banging, giant objects shot into the pile and made it shift and fall and pile up again. Victor's raft was thrown—shot, really—onto that violent landscape, too. Still holding his tin raft, he started jumping desperately. A tree shot out of the water: he dodged it. A girder fell from the pile: he jumped it. He jumped and dodged missile after missile, from every direction, until he saw a whole freight car rear up over his head and he knew he couldn't jump that. It came down at him.

But just as the freight car was about to land, the three-story brick building, against which all this stuff had been pounding, abruptly gave way, and Victor's raft shot from under the falling freight car. He was out in open water again, still racing along.

CHAPTER 9

THE NIGHT OF THE
JOHNSTOWN FLOOD

Everywhere in Johnstown that afternoon, some version of the nightmare was playing out. Compared to Mineral Point or even Woodvale, there was a lot of Johnstown to play with: this was an industrial city with lots of buildings and thousands of people, and it was at the bottom, on that plain in a deep hole. The waters having arrived, they wouldn't fall away as fast as they had from the towns above, and the surging destruction went on and on, giving plenty of time for a multitude of horrors to occur.

At first things went pretty fast. When the water hit Johnstown, it had taken about an hour to travel the fourteen valley miles from the former dam site. But that doesn't mean it hit Johnstown at a speed of 14 miles per hour. The great, tumbling, mountainous wave had achieved wildly varying velocities on its journey. Its huge freight of trees, houses, earth, and wreckage had caused the long pause up at the viaduct, and there were other such temporary halts, where the water would be driven back, spray would shoot fifty feet into the air, the surface of the water would boil furiously. Then whatever blockage would let go, and the mass would roar on down the valley, its pent-up fury released at amazingly high speed. Meanwhile, the narrowing and widening of the valley's turns allowed for speed and relative flattening at some points of the big, overflown channel, and rising and relative slowing at others.

Arriving at Johnstown, the wave hit the city with full force at high speed right in the center of town, where the streets had already flooded deeper than they'd ever flooded before, eight to ten feet in some places. Unlike Rev. Beale's house, most of the big brick buildings in that part of town subsided in an instant, slamming into one another in the water, dropping all their brick to the bottom, throwing people who had been cowering in attics under the water and under the wreckage, the houses' frames and roofs soon bobbing to the surface and racing around.

That was the wave's first move: for about three square miles around the banks of the Conemaugh River, with its banks already deeply drowned, everything simply disappeared. Mercifully, perhaps, a lot of people never even knew what hit them. Knocking down pretty much everything in the center of town took the wave about five minutes.

But now the wave started to divide. It found multiple paths. One part of the flood, a smaller part, broke through a row of brick buildings, and like something hunting, actually seeking to injure and destroy, it ran through that opening and hurled itself as hard as it could against a big Methodist church. The church held up—that's where Victor Heiser did his jumping and dodging routine—so the wave split itself again, with a new section whirlpooling across what had been the city's small, rectangular Central Park. This section of flood spread itself out here, obliterated an entire street of fancy homes, and demolished buildings on Market Street.

Meanwhile, the biggest part of the wave ran, barely impeded, all the way down to the point at the confluence of the Little Conemaugh and the Stony Creek, shot right across the point, went over the river confluence, splashed full force up against the mountain on the other side, and was thrown down the Conemaugh to encounter the arched, stone railroad bridge spanning the river below the point.

This bridge was a substantial creation. George Johnston's train was crossing it like a bat out of hell. Accommodating four tracks, it was fifty feet wide on top, thirty-two feet above the normal waterline, and featured seven squat arches. As the wave met the bridge, tons of wreckage quickly clogged up the low arches, and as with the viaduct earlier in the trip, the wreckage now made a barrier against these archways. With no outlet, the water began to pile up behind the bridge, held by an angle formed by the bridge and the high hill across the river from Johnstown. What one might have hoped—that the water, having achieved maximum destruction, might now move on into the Conemaugh, its force still flooding, yet its energy expended, and travel down the valley to the broader plain—couldn't happen.

And when the other two sections of wave, circling and circling, rejoined the wave piling up against the wreckage-blocked railroad bridge, the whole floodplain where Johnstown sat turned into one huge whirlpool. It would have been far better if the railroad bridge had fallen. It didn't, and with all that debris, piling in some places up to fifty feet above the level of the track, this new, crazy dam was lifting the water even higher than before.

The blockage also caused a countercurrent. Water whirling into the dam of railroad bridge and wreckage began backing hard up the Stony Creek the other way. That action started sending the fallen buildings, which the wave had brought all the way down to the bridge, all the way back the way they'd come, and in new directions, too, crashing them into other buildings now on their way downstream. Those people still alive, riding on roofs and other floats, found themselves racing down to the point and the bridge, then going back up on the countercurrent, and then, when a piece of the wreckage finally did break at the bridge, and the water began sluicing through, being sucked back down toward the bridge.

Even that action wasn't a simple, if terrifying, ebb and flow: it was a twisting, swirling, bucking action, with the banging and crashing of the wreckage dashing people's bodies to death while others watched. By now a total of about 1,600 buildings had been added to the weird soup of railroad cars, bridges, trees, carcasses, barbed wire, and timber. The water had also ripped out riverbed, pulling huge rocks and dropping and piling them at random. So the water and wreckage kept piling higher against the bridge and the hill across the river, making a deep, violent new lake out of Johnstown. As with the original dam at the South Fork Club, this water must soon top the strange dam that it was even then creating out of railroad bridge and wreckage.

And it did. Having risen higher than the bridge and even higher than all the wreckage, the water crossed over the bridge. The railway embankment on the hillside across from town therefore began to subside. Buildings, having been held back by the big stone bridge obstruction, shot through the break and started taking out houses on Iron Street, in Millville, next to the embankment, and flooding Cambria City.

At last night fell. But night brought Johnstown no relief, only new horror.

◆ ◆ ◆

First flood. Then fire.

That afternoon Charlie Dick had given up playing like a kid in the water and insisted, in the face of his wife's derision, that they get the family into the house of a friend on higher ground. Hoping to warm up and get over his chill, Charlie had fallen asleep there.

The next thing he knew, his friend was shaking him and pulling him out of bed. "The dam's burst!" the friend was yelling.

Charlie leaped up, pulled on his rain pants, and ran down the stairs. The water in this house on higher ground was three feet high and rising. Charlie grabbed his youngest child, told his wife to bring the other two, and led them out into the street, pushing through water up to their waists. They were heading for the hillside, and they joined that great throng of people belatedly heading there, too.

Like others on the hill, Charlie stared aghast at the destruction as it occurred.

He'd always thought he had some nerve, yet his heart broke as he watched. For him, it was the heads of the people caught in the wave, the faces of people at the windows. Charlie watched as the huge, clogged flow of houses and people poured steadily toward the railroad bridge and stopped, as the mass piled high against the stone arches. At least, he thought, he'd now seen the worst.

He was wrong.

As the water piled all that wreckage against the arches of the bridge, it was bringing not only buildings and timber and rail and railcars, but also working stoves, working coal-fired steam engines and boilers, working kerosene and oil lamps, molten iron and steel, pieces of natural gas lines, and railcars full of lime. The wreckage was actually a highly combustible tinder-box, with chemical reactions too powerful, despite all the water, to be dampened.

Now the fire started.

While water itself can't burn, it can offer a chemical fire a helpful platform and a ready means of expanding. As later generations would come to understand from the unhappy experience of oil rig fires out to sea, water can amplify fire. Some fires can even burn underwater.

That's why, on the smaller scale of home and restaurant kitchens, we're advised never to throw water on a grease fire. Oil and water, everybody knows, don't mix. When water is poured on an oil fire, it doesn't drown the fire but falls straight through it, landing below the fire and causing the fire to leap high and wide, possibly catching new fuel.

And while the immense water of the Johnstown Flood would have put out any fire burning solely in the timber and frame houses of the wreckage, it couldn't put out either the burning liquid that came with oil and gas, or the burning metal that was molten steel and iron; the coal, for its part, gave those fires more fuel. Oils, lighter than water, skimmed the water's surface and gave both field and fuel for the fire; water interacting with molten metal only fed the fire, and wherever the wreckage piled above the waterline, the fire burned like a pyre, wet but nowhere near saturated enough to resist serving as food for a blaze with this much chemical prompting.

And so even before dark, fire was turning the whirlpooling action at the bridge, where so many people were still alive on their floats, or lay stranded, injured, on the wreckage, into a great tub of leaping orange flame. Buildings and people were still being swept down toward the bridge, some for the second time. And now they were swept into a great sea of fire.

Charlie Dick watched from the hillside as the flames leaped and spread and people wedged in the wreckage and stuck in the floating houses began to burn. As the victims yelled for help, those on the hillsides stood paralyzed with terror. There was nothing they could do but watch or turn away, still hearing the futile screams.

Charlie saw men and women in the fire hug each other goodbye. As he listened to the shrieks, he prayed only for it to go fast, for the people's suffering to end soon.

It didn't go fast, and Miss Minnie Chambers, who had been riding the waves in a boxcar, found herself forced to hear the screams of the dying late into the night. One man finally had managed to swim through the doorway and into the boxcar; together they'd taken the wild trip down toward the bridge. Minnie watched dead people float past, others still alive on roofs and floors, some kneeling with clasped hands, as if praying, others just white with despair.

Suddenly the boxcar spun around. Looking out the doorway, Minnie and her companion could see the stone arches of the railroad bridge, just ahead. They were going to hit that dam of wreckage at full speed.

The man yelled to Minnie. They must jump into the water or be dashed to pieces on the wreck.

Minnie didn't want to leave the boxcar. So as they hurtled toward the bridge, the man grabbed a plank lying on the car's floor and jumped. Minnie watched: the circling current grabbed the plank out of the man's grasp, and the plank smacked him on the head. He threw up his arms.

The man sank. Minnie covered her face as her car veered toward one of the stone arches of the bridge.

When the boxcar struck the arch, the struck side fell right off the car. Water rushed in. The car stayed there, but Minnie was again swept up and away, she couldn't have said where. In the water, she banged hard against something, and her hand went out automatically.

She grabbed the thing. She clutched it, not knowing what it was. She pulled her head above the water and kept holding on, while the water swirled about her and tried to drag her away.

She seemed to be clutching a big iron pipe sticking up through the water. The pipe wasn't moving, and Minnie just kept holding on to it, resisting the pull of the water as night fell, and an orange glow filled the dark sky, and she grew colder and colder, holding the pipe.

With the city's electric light service demolished, the only light in that flooded, turbulent, deadly whirlpool, once Johnstown, came from the orange glow, at times bright enough to read by, a new disaster burning hundreds of helpless, shrieking people alive. Thick, foul smoke from the fire soon began to surround Minnie, but she held on. The screams and groans of those dying in the fire surrounded her, too, late into the night. Minnie held on.

The company-store clerk Thomas Magee was still on the store's third floor, still guarding the $12,000 in folding money he'd saved from the safe, his team of clerks with him: Michael, Frank, Chris, Joseph, and another Frank. This team thought the building seemed surprisingly solid. The third floor wasn't the worst place they could be.

So Magee and the boys started doing everything they could

to save everyone they could. Taking shifts and dividing up tasks—at least one person always guarded the money—they spent Friday night helping. When they saw one man who was trying to jump to their window fall into the water, they reached for him as a team, and because he was a strong swimmer, they got him through the window to relative safety. After that, they started throwing out ropes to snag and pull in boards and logs. With these, they made a raft, tying the timbers together with ropes. Tying the raft securely within the building on long ropes, they floated it out into the flood and used it to help people climb inside.

As the long night went on, eighteen people, including six children, came out of the water into the third floor of the company store. And where they couldn't reach people without getting swamped themselves, Magee's team used the raft to float the bread and meat they'd wrapped in canvas and oilcloth out to people passing by. Some packages they just tossed.

They fed maybe one hundred people that way. And they kept guarding the cash.

The Reverend Beale, Captain Hart, and the other adults from the Beale attic, having gotten everybody down onto the nearby rooftop, including Guess the dog, were trying to lead their families in picking their way from debris to debris in search of safety. They had a better chance now. The Beale house was near the center of town, where the wall of water had hit first, and with the violent action mounting down at the railroad bridge, some of the wreckage had settled down here—though it was still moving precariously—and some of the water had at least somewhat calmed.

The Beales and their friends' first objective was the reverend's church, only about fifty feet from the house. But there was no wreckage to climb that way, only swirling water, so they sought another direction.

They thought of Alma Hall. The biggest building in the city, four high-ceilinged stories, wasn't far away. And in that direction, mountains of wreckage were piled high above the water.

So walking and jumping from moving object to moving object, from rooftop to boxcar to tree, and onward, they tried for Alma Hall. Sometimes some members of the group couldn't see one another behind roofs. Sometimes they had to cross the wild water on loose boards used as bridges. Suddenly one of them slipped, a young woman, going right into the water: they could see only her hair, floating on the water's surface. The group scrambled to pull her up onto some floating timbers.

They all did arrive at the second floor of Alma Hall, just before dark, with the new horror of the fire just beginning to glow in the distance. The Beale group was hardly alone here. The place had drawn a highly varied assortment of people from all over town.

One of them was James Walters, a lawyer, whose arrival here had been truly bizarre. Walters had an office on the second floor of Alma Hall—yet he'd been at home on Walnut Street when the wave hit. When the rest of his family floated away on a piece of roof, Walters had found himself riding the flood alone, inside his house. When the house struck Alma Hall, it dumped him into his own office.

By the time the Beale party arrived, about three hundred people were spread throughout the second, third, and fourth floors of the big building: Gentiles and Jews, as the Reverend Beale noticed, and Catholics and Protestants—and Africans, as Beale separately distinguished them. And Chinamen, too.

The men now held a meeting on a staircase. Probably, that

is, the white men did, of the better-off class. In the strange Alma Hall government they quickly formed, the offices were assigned to Walters, director of the building; and to Beale, Beale's friend Captain Hart, and a Dr. Matthews, each monitoring one of the floors in use.

Mindful now of the awful fire burning nearby, this self-appointed governing body outlawed turning on the lights: the building had gaslight, and they feared it might explode. For the same reason, all liquor was to be handed over to the monitors for safekeeping.

So the pocket flasks came out and were taken charge of. Rev. Beale then made the rounds of the whole hall, in response to a motion of the directorate, offering a prayer in each room. Walters went with Beale and suggested that all should bow their heads during the prayer, and everybody did regardless, Beale was impressed to note, of race and faith.

In the darkness that night, the inhabitants of Alma Hall heard the waters roar and the screams of people dying nearby. Children in the dark hall cried for food, dogs howled, and the adults prayed, sobbed, and moaned, many with bad injuries, many missing children or parents.

Smells were sickening, stifling. The fire lit up the sky. And contractors and builders in the hall were sure that Alma Hall, too, was just about to go.

Papa," Gertrude said.

The little girl was alone in the dark and the water now, spitting, gasping, tossing her head.

She saw a glimmer of light. It was high above her head.

She moved toward it and started climbing. Gertrude had recently become a good climber of apple trees. The light began to gleam while she scrambled upward along something rough, and upward some more, toward what was turning into a bright hole leading outdoors.

She crawled through it and sat down in the rain. She was on something like a raft, floating and tipping and circling as it carried her over fast, roaring waters. A muddy, soaked mattress was on the raft, too, with sheets and a quilt.

She held on to the bouncing raft and tried to look around, but her raft kept tilting back and forth. The water was splashing, frothing, tossing between all the hillsides. Everything looked like taffy boiling in a pot.

The tilting made the little girl keep jumping. She sprang about like a cat, trying to keep the raft balanced.

Then Gertrude's ring slipped off her wet finger. It dropped on a flat board going by in the water. A gold-band ring, with little blue enameled forget-me-nots. Gertrude loved it.

The ring sat on the board, which was heading straight for the hole she'd just crawled out of. She jumped to the end of her raft and grabbed for the ring.

She got it. Pleased, she put it back on her finger. She sat there admiring it.

One end of her raft shot into the air. It had collided with something, and Gertrude held on, gripping, riding out the collision. Looking back, she saw she'd bumped into a horse. She watched as a big tree pulled the horse into its moving branches. The horse was dead, but caught in the tree branches its body rolled slowly upright and it came down the flood after her. It was rocking back and forth like a horse on a merry-go-round.

Now Gertrude was scared. She was shivering.

She lay on the soaked, muddy mattress. The quilt belonged on her mother's bed. It was handmade—Mama had told her that—and showed a pageant scene, and Gertrude liked hearing her mother tell her about it.

She thought about that now, looking at the quilt. It was getting dark. She closed her eyes.

But the rain kept hitting her face, and the water was loud. She covered her face with her hands and started crying. She was wearing, she now realized, only her underclothes.

Then she got up on her knees. She prayed aloud, the German prayers Mama had taught her. People went flowing past the little girl, living and dead, but all the animals she saw were dead, their hair matted and soaked, their eyes just staring.

In the distance she saw tombstones, decorated with flags.

Near Gertrude a small white house came floating along. A man was straddling its roof, holding on to its chimney. Gertrude called out to him. She begged and demanded that he save her.

But he didn't see her, or else he was just ignoring her.

Then he was gone.

"You terrible man," Gertrude said aloud, once he was well out of earshot. "I'll never help *you*."

She stayed in her kneeling position—upright, the way Mama had taught the children to pray. "Let the little children come unto me, for of such is the Kingdom of Heaven": she knew that saying, from the Bible, because Mama had told it to her. And she thought of the catchy Salvation Army song she'd once heard:

"You must be a lover, a lover of the Lord . . ."

"I am a lover of the Lord," Gertrude protested. She said it as loud as she could.

". . . or you can't go to heaven when you die . . ."

Darkness had fallen.

CHAPTER 10

ALONE IN THE WORLD

It turned out that what Miss Minnie Chambers was holding on to, all through the night, was a pipe that came up through a roof of one of the Cambria Iron Works buildings. The pipe didn't fail her, all that awful night. The building stayed put. Minnie held on.

When the day dawned, gray and misty yet without rain, the waters were down, leaving Minnie on top of a huge pile of wreckage. She was freezing cold, and stuck up there, badly bruised. Some men came by and helped her climb down. They took her to some friends in the town of Prospect, and aside from those bruises, and despite her amazing ride, Minnie Chambers had no injuries at all.

Mrs. Stevenson and her daughters, after breaking the plaster ceiling on their second floor and climbing up through the attic floor, had ridden the waves for a long time until suddenly the house landed on dry ground and wreckage in front of a Dr. Walters's house on Vine Street. They left the attic, went into the house, and made it through the night.

◆ ◆ ◆

In the end, Alma Hall didn't go down. At the break of day, Rev. Beale looked out the door of the building and saw that the water had subsided. The flood wasn't gone, but it was down and falling farther.

Some in the building had suffered broken bones and other injuries, but most were able to get out of Alma Hall. Many had been separated from family. Some of them still had hopes of seeing their loved ones alive again. Others had certain knowledge that they never would: they'd seen their loved ones die. Some of those survivors, overcome with grief, as well as with horror at the ways in which those they loved had met their ends, felt envy for the dead.

But all of the survivors, whether rescued as family groups or left alone, whether sure of tragedy or hopeful for reunion, felt battered, hungry, exhausted, and stunned. Small children hadn't been fed in sixteen hours or more. Stumbling awkwardly through a window facing onto Main Street, nearly three hundred overnight residents of Alma Hall climbed onto the wreckage around the building and took a long, hard look at their city.

What they saw was just as hard to take in as the great wall of water. There simply was no Johnstown. The wreckage they were standing on just went on and on.

In the gray, dank weather—the rain had stopped, for now, but the morning sky was dark—wherever a building stood, it stood out, often crooked and leaning, lonely against the sky or the slopes of the surrounding hills. All else was a mountainous chaos of debris, piled in some places as high as those few buildings that still stood. Many survivors had lost people. Everyone had lost almost everything else.

And there was no sound, not really. Johnstown was normally a loud place. That noise came from industry. There was no industry today, and the quiet was eerie, a sign not of peace but of incalculable death and destruction.

The Reverend Beale and his family started into the wreckage, heading for a hill at the foot of the steeply climbing Frankstown Road, past the edge of the flood's path, where other survivors were gathering. The Beales and others from Alma Hall crawled and picked their ways through a landscape. This landscape was made of freight cars, houses, bridges, trees, furniture. And dead bodies.

When they arrived at the foot of steep Frankstown Road, they found about three thousand people gathered, just as stunned and grief-stricken and hungry and exhausted as they: every age, every kind of person, cold, with little clothing, stunned. Everyone was injured or bereaved or simply horrified to the point of insanity or all of the above. From here you could see that fires were still burning in various parts of the wreckage.

Still, of these three thousand people, not one adult was crying. Everybody was keeping a face so tight, Beale thought, it was as if their heads were held by a vise.

Beale's first thought was for relief of his own family, so he walked them up the steep hill to the home of a friend who lived high enough to have escaped disaster. He left his wife and children there, for care and food and sleep, and went back down to the gathered survivors to see what he could do to help bring order to this horrible confusion.

First, the group had to find shelter for the women and children, as Beale had done for his own. People up on the hills were offering help, and as quickly as possible, shelter was found there. Still, there weren't many people living on the hills, and the houses up there were poor and small. Thirteen families went into one house. Nineteen, with nothing to change into, went to sleep on the floor of another.

The survivors below, mostly the men now, kept gazing back at the site of the former city. St. John's Catholic Church had burned in the fire: it was a hulk, still smoking. Other church spires were simply gone. Almost nobody in this group any longer owned a house, a bed, clothes. They stood there hatless—a bizarre experience in 1889—many without shoes, many nearly naked, many physically battered. There were Civil War veterans in the crowd who said they'd never seen anything like it.

Beale, as overcome as anyone else, had nothing to compare this situation to, but he was getting a worrying idea. He recalled reports from the Franco-Prussian War, a little under twenty years before: the day the Prussians took Paris, it was said, that city was subjected to thieving and looting and anarchic behavior on the grandest scale.

Beale saw a little boy on horseback. He gave the boy some money and told him to find a working telegraph office. Governor James Beaver had to send military assistance right away.

So the monster had moved on. Soon it ceased to exist.

Literally. All of that water went somewhere, of course, but the monster was nowhere. The monster had been created by specific conditions. Because they involved motion, and the release of energy, it couldn't live long. Now all that gave testimony, or even size and shape, to the monster's brief and terrible career was the scope of destruction it had left behind. From the scale of destruction, some of the monster's features can be calculated.

The water's volume per second, as it flowed through the former South Fork dam, might be called equal to the volume per second going over Niagara Falls. Yet sheer volume unleashed wasn't the only thing that caused the incredible destruction, not only to Johnstown but to all the towns on the way to Johnstown, and after Johnstown, once the water made it past the railroad bridge, to towns on the Conemaugh like Cambria City, all the way west to Bolivar, and southward, on the Stony Creek, to the area below Moxham. Mass is one thing, and the water picked up many more tons of mass. Speed is another, varying as the water came down. If 20 million tons of water fell 404 feet, the real enormity of the phenomenon came from the fact that all of its energy had to be fully released before the water could come to rest. Some part of that energy was used up in resisting the friction between the water and rocky floor and sides of the valley.

But the balance of the energy could, in a sense, actually be witnessed, or at least reconstructed mentally from what could be witnessed, in the aftermath. That energy, now expended, was visible in what the whole valley, the city of Johnstown, and some of the towns just west of Johnstown looked like now. The monster that no longer existed had been made of the release of its own energy, and when the stunned, half-naked, chilled, and ravenous citizens of Johnstown surveyed the scene, the monster was already long gone.

Even as the fire at the bridge raged and devoured people alive, the dam of debris that the water had made at the railroad bridge began to fail. Water ate away at the embankment on the side across from town, deeply enough to give channels for escape. That sent water shooting through to those corporate-named towns Millville and Cambria City. Much of the energy had been spent by then, and the bridge slowed the water down, but not enough to keep the flood from wreaking havoc on those two towns. Their fate too was part of the Johnstown Flood.

Below those towns, however, the water could begin to spread and come to some kind of rest, a broad, deep flood, not a force of outrageous violence. The Conemaugh Valley widens there, and for a time it straightens, losing its characteristic tight bends, and the plain is flat, inviting the water to stop hurtling and fighting and piling up and rolling, inviting it to spread out and lie down. Also, totally destroying Johnstown and Cambria City and other towns nearby had taken some of the fight out of the thing. So here it finally died, scattering bodies and debris as far as Bolivar, yet with no further ability to rip and tear and beat and kick things down.

This was only, after all, a lot of H_2O. It had formed no intention of becoming a monster. The 20 million tons that had so recently been stopped by a sporting club dam went, now, where it had been heading in the first place: down the valley of the Conemaugh, into the Allegheny, into the Ohio, into the Mississippi, into the Gulf of Mexico.

The crew of clerks up on the Cambria company store's third floor, having rescued and fed so many people that day and night, were rescued themselves, by boat, on Saturday morning, and they went around rescuing still more people. Thomas Magee still held the company cash he had saved from the safe.

Magee would hold on to the money till Monday. Then he would return it, fully intact, to a superior.

Victor Heiser, who had leaped and jumped and climbed and swung, timing every move, through such an incredible number of adroit escapes and dumb-luck near misses, had seen his parents disappear when his house was crushed, but Victor was sixteen, and an only child. Now that it was over, he thought they might be alive, that there might be news of them, that somewhere in the astonishing wreckage of Johnstown he might rejoin them.

When he'd been swept from under the falling freight car at the last minute into open water, Victor had known generally where he was: coasting through water many feet over Central Park. He'd also begun to feel his second-to-second danger less intensely, and with time to look up he could see his neighbor, Mrs. Fenn, riding the wave. She was straddling a tar barrel, rocking and rolling, almost going under on each roll, covered with tar from the barrel.

Victor didn't yet know that Mrs. Fenn had lost her husband and all of her seven children. He wanted to help her, but there was nothing he could do. He was being swept down toward the railroad bridge.

Yet unlike so many unlucky others, he was thrown back by the countercurrent, up into a backwash along the hillside, and he found himself slowing down. Determined to take quick advantage of the fact, and seeing the roof of a submerged but still-standing house near the edge of the hill, he jumped from his tin float. He landed on the roof. A group was sitting up there, stranded.

That's when Victor Heiser got a strange idea. He might live.

He looked at his watch. It wasn't 4:30 yet. He'd been fighting for his life for only ten minutes.

Many others would face horror and death to come, in the fires and in the water, during the long night of terror. But for Victor the worst really was over. He didn't know it yet. The group on the roof was cut off by ebbing, flowing, cycling water from the safety of the hillside, so near and yet so far, and houses nearby kept going down into the water. They spent the rest of the day trying to help others get onto their roof, and after a while nineteen refugees sat there. When it got dark they decided to climb down into the attic of the house and get out of the rain.

That's how they spent the long night: hearing, like the denizens of Alma Hall, awful sounds—in this case the occasional *whoosh* that told them yet another building was going into the water. As exhausted as he was, Victor didn't sleep. To him, this waiting seemed almost worse than anything he'd been through after getting hit by the wave. If the attic collapsed, they'd all die, trapped. All of his effort and luck would be for nothing. The suspense felt deadlier than the fight.

But at dawn, Victor climbed down, like every other survivor, onto the amazing expanse of wreckage. The water was down to flood level now, submerging only the first floor of the house, and the debris, no longer moving, was navigable. Victor found his way over piles. He waded through shallow water. He used debris to float across the deeper water. Here and there he actually found some solid ground.

He was looking for his parents. In that quest he encountered other survivors, and everybody seemed to be doing the same thing: looking. He saw dead bodies, and he saw people pulling dead bodies out of the pile. If his parents were dead, he hoped at least to find out for sure, to find their bodies. He hoped to bury them decently.

Victor was nearing the railroad bridge now, on the far side, where the eaten-away embankment had finally let the water out. During the night, Victor hadn't known about the fires. This morning he saw: flames here at the bridge were still leaping, and along with horribly burned dead bodies, many people, injured yet alive, were still stuck on the pile, unable to escape as the flames burned their way nearer and nearer. The trapped were shrieking in terror—men, women, and children—as the flames approached them.

There was, Victor saw, more agony and death to come this morning. When he saw rescue teams climbing the pile, trying to reach and free the trapped victims, he jumped in to help.

For hours that day, Victor Heiser and the others tried to save people from what became, as the flames spread and began consuming the pile, a towering, roaring funeral pyre. They did save some. But they had no axes, no tools at all, and they had to keep ducking back out of the fire or else burn up themselves.

So while Victor helped save people, he also had to watch, helpless, as many burned to death before him. Some of those dying in agony were people he recognized.

Finally he ran into a friend from outside Johnstown. The friend's house hadn't been affected by the flood, and Victor went home with him. The family took the youngster in, fed him, gave him a bed.

Now Victor knew. He was alone in the world.

A huge painter and paperhanger named Maxwell McCachren had not only survived that night but had also performed a feat that would make him forever famous in Johnstown. Just as darkness had fallen, he'd been riding the wave on a big, pitching roof, crowded with maybe twenty other people, everybody holding on for dear life and certain that at any moment they'd be tossed into the water. Then they saw something astonishing.

A little girl, maybe five or six. She was upright on her knees on a raft, with a mattress and quilt. She was wearing only underclothes.

They could see that she saw them, too. The girl started shouting for help. She wouldn't stop.

But really there wasn't anything they could do. The girl was too far away across the churning, choppy water, and anyone who let go of the roof, even for a second, was likely to slide into the flood.

Then some in the group saw Max McCachren moving toward the edge of the roof.

Max was a big, strong man of thirty-eight, and an emotional man, and the others could see what he was thinking. It was crazy: if he went after the little girl, he would surely drown or be bashed to death. Max was married. He had fifteen children. Some of his fellow riders grabbed at Max, demanding that he stop. "I'm going over to save that baby!" Max said. "Do you people think an angel from heaven is coming down to help you? God helps them that help themselves. . . ."

Gertrude, having stopped announcing that she was a lover of the Lord, was watching this argument on the rooftop across the water and keeping up her shouting at the people. She cried loud tears of rage and fear; she demanded, she pleaded. The roof kept throwing those people around, but that meant nothing to her. She just kept shouting at them, crying, demanding, begging.

Now she saw the big man on the edge of the roof shake off the people trying to hold him.

He jumped off the roof into the violent tide. He disappeared under the water.

The man's head bobbed up. He was swimming, and he was looking at Gertrude.

She kept up her yelling. At top volume, she cried and begged. The man kept disappearing underwater. Then he would come back up. Sometimes she couldn't see him at all. But she just kept yelling, and somehow he seemed to be managing to catch up to her circling, moving raft.

As Gertrude watched the man swim, she stopped yelling. This man was going to save her, she decided.

* ◆ ◆

Max had reached the raft. He grabbed its side and pulled himself up across the mattress and lifted the girl. She put her arms around his neck. She had an amazingly powerful grip. His heart swelled with joy.

The roof he'd come from had now bobbed far away. Max held the girl while they flowed downstream on their circling, tipping, precarious raft, while people cried and moaned in the dark and debris boomed and banged around them.

Max thought the girl might be about to strangle him. Yet her grip gave him a warm feeling, too.

Then he saw that they were approaching the stone-arched railroad bridge. He saw fire raging, smelled a chemical smoke, heard screams. He was taking the little girl straight into another kind of hell.

From the window of a small house up on the hill, two men were poking long poles down into the water to guide people in. One was a white man, the other a black man. Max leaned and kicked, still holding the girl, trying to get the raft to veer toward that house, but he couldn't do it: the men in the window were still maybe fifteen feet away, and Max couldn't perfectly control the flow. He and the girl were going to float past the house, right down to the fire at the bridge.

The men in the window yelled to him, "Throw that baby!"

Max wasn't sure. Not about his throwing ability. He could throw.

He yelled back, "Do you think you can catch her?"

"We can try!"

What else could they say? Or do? From the window, the white man leaned far out, so far he seemed to be about to topple into the water. The black man got behind him, bent down, and held him by the knees so he wouldn't fall.

Max broke the little girl's grip.

◆ ◆ ◆

Gertrude Quinn found herself flying upward through the air in the rain over the flowing water. The man leaning out the window reached way out.

He caught her, and he held on. The other man pulled. They all tumbled into the room.

The man on Gertrude's raft sailed off into the night.

Later that night, Gertrude was being carried around the hillside in a dry blanket, her wet clothes gone and so bundled up she couldn't see out. "What have you got there?" she heard some men say.

"A little girl we rescued," said her carrier.

"Let's have a look."

The blanket was pulled partly away, and Gertrude looked at faces peering down at her, staring, actually, and they scared her. One man squinting at her had a red face and turned-up nose. She thought she'd better reach up and yank that nose.

She didn't get the chance. "Don't know her," they said and the blanket went back over her face. Now her carrier was climbing eight steps to a porch. They'd arrived at a long, three-story frame building, with six apartments: worker housing on the hillside. Gertrude was unbundled inside the house and placed in a woman's lap. The woman held her close, rocked her, and sent one of her many children to the attic to get some red flannels, just packed away for summer.

"This poor little freezing child," the woman said. She sent an older child to fill some Mason jars with hot water to warm Gertrude up.

Gertrude hated red flannels: they itched. But the kitchen stove was warm, and yellow lamplight seemed, after where she'd been, to shine like a diamond.

She was scared, too. Tenants from the other apartments had come in, as well as people off the street, and everybody was gathered around looking at her with their kind faces, asking her question after question.

She decided the best thing would be to not answer them. She fell silent.

She shared a bed upstairs with three women, also taken in from the flood. Pretending to be asleep, she peeked at these women out of the corner of her eye. They were sisters, and they kept getting up and going over to the window, looking out, and gasping.

They didn't want her to hear, so they whispered. "Frightful . . . terrible . . . ghastly."

When the sisters finally fell asleep, Gertrude slipped out of bed.

She went to the window. She looked out.

Below the hill, only water spread everywhere, and firelight flickering here and there reflected on the flowing surface like an illustration plate of ships burning after a battle at sea.

Gertrude's new strategy was silence, and she kept it up the next morning.

But the three sisters—their name was Bowser—had a theory. They thought they recognized her.

This small house on the hillside was full of flood refugees, and the household was up before daybreak, trying to cope, in horror like everybody else, with the wreckage of their city and the loss of their loved ones. At about 5:00 that morning on the crowded porch, the sisters asked, "Aren't you little Gertrude Quinn?"

Gertrude didn't respond. She was wearing clothes belonging to one of the many daughters of the house, and mismatched shoes, one too large and one too small. She hadn't had a bath. She sat there looking at everything, her hair full of dried mud, her face dirty.

She also knew by now the name of the woman who had warmed her up the night before: Mrs. Metz, already in the kitchen this morning, cooking for her crowd of refugees. Now Mrs. Metz, having overheard, came out on the porch and widened her eyes at the Bowser sisters.

"Surely not one of the Quinns of Quinn's Store?" Mrs. Metz said. They widened their eyes back and nodded.

This might be a good thing, Gertrude figured. The store was well-known. Maybe, she thought, since she came from a respected family, she would be invited to live with the Metzes. She was careful to act as if she couldn't even hear them talking.

"She does look like the little Quinn girl," one of them said, sotto voce.

"Maybe," said another, "if we could get some of the sticks and mud out of her hair she would look more human."

So they'd seen her before. At the store. Evidently she was hard to recognize now.

And they kept asking if she knew her name. Did she know her father's name? But Gertrude was thinking she might be the only person in her family left alive. She was filled with a new kind of dread.

Suddenly the Bowser sisters called out to a woman walking by across the street from the Metzes' porch.

"Mrs. Foster," they called, "do come over here and see if you can identify this child. We think it's your niece, Gertrude Quinn."

Mrs. Foster crossed the street in a hurry. Gertrude just sat there.

◆ ◆ ◆

James Quinn was still on the hillside. With Helen, Rosemary, and the baby, he'd spent the night in the same way that others who lived through the flood had spent it: in a horror of over-whelming grief and loss and terror, facing or hiding from the death and destruction raging below.

James was in despair. For in the end, as he'd watched from the hillside, having barely escaped the huge wave himself, he'd seen his whole house go all the way down into the water. He knew his little Gertrude was in that house.

On the hill, James had found the home of Mr. and Mrs. Leis—another of the poorer hill dwellers. Mrs. Leis worked at the Quinn store when she could spare a minute from her eight children—and the Leises had somehow found the room to take them all in. James spent the night on the porch, pacing. Back and forth he went, crying for Gertrude.

"My poor little white head," he kept saying, over and over, holding his head in his hands. He could have carried her, as well as the baby. He didn't know why he hadn't.

At one point Helen said, "Father, Vincent is missing, too. You don't say anything about him."

"Vincent is a boy," he said. "He can swim, he had a chance—but my baby had no chance!"

No chance: that's what was destroying him. Also, his last exchange with Gertrude had been a spanking, followed by an angry lecture. If he could ever see his little girl again this side of Kingdom Come, James kept saying, he would never chastise her again, about anything. . . . And he went on pacing.

About 5:30 in the morning, James was at the bottom of the porch, washing his face in a basin of hot water the Leises had provided, when his sister-in-law Barbara Foster appeared, nearly out of breath.

"James!" she said. "Gertrude is alive!"

On the Metz porch, it had taken Barbara Foster a moment of looking closely at that filthy, terrified, staring, strangely dressed girl to see that, yes, this was indeed her little niece. And it was confusing: Barbara could have sworn that, as she'd run for her own third floor the day before, she'd seen James wading into the street with his girls.

Anyway, Barbara had bolted right down the Metzes' porch steps and run all over the hillside looking for James, and now she'd found him and told him the news. He looked up. He put his hands in the air. He sobbed. He begged her not to say it.

"James!"

"I saw the house go down," he said, "chimney and all, and slide under the water. My poor little white head had no chance."

Barbara had to put a stop to this.

"James!" she said. "I've talked to Gertrude. She is frightened nearly out of her senses!"

James Quinn came to life. Barbara turned and ran, and he ran after her.

Gertrude was still sitting on the Metz porch with the Bowser sisters when she saw her aunt running toward them, with Papa close behind, and Helen and Rosemary bringing up the rear, and the little girl hurled herself down the steps, landing on Papa's knee just as he put his foot on the first step, putting her arms around his neck as he hugged her.

Both were crying, and Helen and Rosemary were pushing into the hug, too.

"My poor little sister!" Rosemary cried. "Oh, my poor little sister, we'll never lose you again!"

The crowd on the porch was swelling with people marveling at this reunion. Mrs. Metz offered to take care of Gertrude for as long as Mr. Quinn wanted, and James started to say that would be a wonderful idea: the Quinns had no home of their own now.

But Gertrude started crying. She clung to Papa, and he changed his mind.

"Of course you'll go with us," he said, comforting her. "Even though we have no place to go."

PART III

JUSTICE AND CHARITY

CHAPTER 11

SOME CONVULSION

A<small>NDREW</small> C<small>ARNEGIE WASN'T IN THE AREA WHEN THE</small> S<small>OUTH</small> Fork dam broke and Johnstown and the other towns were wiped out. He wasn't often in the area, and he wasn't even in the country in the spring of 1889.

Carnegie was in Paris, visiting the Universal Exposition. The Eiffel Tower had been erected to serve as an astonishing entrance arch and viewing platform for that fair. At twice the size of the Washington Monument, this tower was now the tallest construction in the world. Machinery Hall, also a feature of the exposition, was the longest interior space in the world, a vaulting series of glass, iron, and steel arches that seemed to go on and on with no evident internal support.

There was also the "Great Model of the Earth," a globe built to exact scale, housed in a high, circular building. You could see the North Pole by riding up in an elevator. The huge globe showed, in a beautiful color scheme, all of the seagoing trade routes and railroad and telegraph lines of the world, all the mines for everything from gold to coal. At the Universal Exposition, the nations and businesses of the world showed off their wares and innovations and cultural practices and military accoutrements. The world seemed to be transforming itself, bringing humankind into what people had begun to call modernity.

It was in Paris that Carnegie got shocking word of the disaster in Johnstown. He'd long had important ties, if sometimes difficult ones, with the Cambria company there. He'd hired Captain Bill Jones away, for one thing, along with Jones's top men, and he'd competed successfully with the late Daniel Morrell. The western Pennsylvania steel world was a very small one. It was Carnegie who had pioneered, too, the love of the Alleghenies as a resort. He knew the Conemaugh Valley well, and this horrible news overwhelmed him.

He wrote to Henry Frick to say that the news from Johnstown had driven everything else from his mind. Almost immediately, Carnegie sent $10,000 for relief of the city. He wouldn't be back from Europe until the fall. He would then visit Johnstown to look at the scene of such overwhelming destruction, and he would provide full funding for the rebuilding of Johnstown's Carnegie Library.

Andrew Carnegie wouldn't, however, express any regret about the failure of the dam at the club of which he was a member. His mind, as usual, was on the higher things.

Nor was Carnegie alone, among the South Fork Club members and big men of Pittsburgh, in not coming to Johnstown in the days and weeks right after the disaster, or saying anything about the dam. Henry Frick and many others would give relief money, and other barons of Pittsburgh industry would organize much of the giving and supply that came in from around the country. Yet members of the South Fork Club were at pains to deny any responsibility for the titanic devastation that occurred when their dam broke.

That refusal began even before they knew the full scale of the disaster. It began even while the devastation of Johnstown was still going on.

Of the South Fork Club members who hadn't been up at the lake when the dam broke, Robert Pitcairn, head of the all-important Pittsburgh division of the Pennsylvania Railroad, was probably the most important, for reasons having little to do with his club membership. And it wasn't because of the South Fork Club that Pitcairn spent the afternoon of May 31 trying to get to Johnstown.

As the top railroad executive in the area, and one of the top railroad executives in the country, Pitcairn was always directly alerted of danger to the tracks and the trains: he charged companies a lot of money to move their freight on his tracks, and when the trains stopped, industry stopped, and money disappeared, and he had to get things moving again fast. In normal times, his days were spent at his gigantic desk in constant communication with many elements of a complicated system. On a day of a record flooding, with track washing out between Lilly, high up on the mountains, and Bolivar, about fifty miles east of Pittsburgh, Pitcairn knew he had to check things out in person.

When Emma Ehrenfeld and the Mineral Point operator had tapped out their telegraph messages that day, warning about the condition of the dam, and sent them down on paper and then again by telegraph to East Conemaugh, and then to Frank Deckert in the Johnstown railroad station, another of their recipients had been Robert Pitcairn. Nobody was more important to alert. Indeed, so important was Pitcairn's role in keeping freight moving that he'd already left Pittsburgh by the time those messages about the dam arrived in his office.

For the dam wasn't on his mind, as he would say later. It was the flooding that concerned Pitcairn. Reports had begun coming in during the night of high water the likes of which had never been seen before. Trains were stopped dead, and by around noon Pitcairn had his private car hitched to the No. 18 passenger train and was riding eastward through driving rain in hopes of locating the source of all the flooding and assess repair. He planned to go all the way up to the station at Lilly, where he presumed the flooding originated, and look in on Johnstown on the way.

The dam up at the club wasn't on Pitcairn's mind because, as he said later, he'd long since satisfied himself on that score. He did have a special relationship with that dam. In the period after the railroad had bought the old reservoir feeder property from the state, and before the company had then sold it to John Reilly, the dam had been among Pitcairn's responsibilities, and the break in 1862, which damaged some railroad bridges, had brought its weakness to his attention. But then the railroad had sold the property to Reilly, from whom the South Fork Club would later buy it. Official responsibility had long since passed out of Pitcairn's purview.

Still, he'd felt some concern. He'd heard that Reilly had removed the discharge pipes and sold the metal—heard that selling the metal had been Reilly's real purpose in buying the property. When Benjamin Ruff had started making plans to rebuild the dam, Pitcairn went up to the property to look things over and discuss the situation. He had a lot of confidence in Ruff. The man had supervised much construction in his day, and by his own report knew more about dam construction than any certified engineer.

So while Pitcairn had paid a number of visits back then, and had sent railroad supervisors and others up to observe as well, he'd become convinced that aside from an early concern about leakage from its bottom, the dam was fine. Ruff had promised to repair that one area of concern, and Pitcairn sent his supervisors to report on whether Ruff had in fact done so, and the report was that Ruff had. Since then, Pitcairn said later, he'd even kept his people checking the dam constantly, at least once a month, sometimes twice. Sometime every year, therefore, he would receive a report from one of his inspectors or another warning him that the dam was damaged or about to break. The reports always turned out to be false.

So if anything, too much vigilance was in play, as far as Pitcairn was concerned. For even if the dam did one day break, he couldn't imagine it doing very much damage. Today, with rain beating down on his private car, Pitcairn had no concern about the South Fork dam.

The flooding, however, was impressing him as he rode east along the rivers. It was soon clear that he wouldn't be able to get to Lilly. Indeed, he soon found he couldn't even get to Johnstown. At Bolivar, between Pittsburgh and Johnstown, he saw water much higher than he'd ever seen before. The train proceeded slowly eastward. At Lockport, Pitcairn saw muddy water as high as the north track. Unheard-of.

Then, only about four miles west of Johnstown, the train stopped. Pitcairn left his car and went up the telegraph tower to see if he could find out what was wrong up the line, but there was already no getting through: wires were down to Johnstown and many points beyond. Without clear information, the train operator couldn't let the train go ahead.

Pitcairn wasn't just a railroad executive. He was a railroad executive because he was a railroad man, the region's top conductor, and he decided to take it on his own responsibility, as head of division. He would go ahead, carefully, and find whatever the trouble was.

But then, still up in the tower, he saw the debris.

Rushing down the river past the track, this seemed an unusual kind of drift, all wood, broken in very small pieces. Meanwhile the telegraph poles started coming down. The tower itself seemed ready to fall.

Then Pitcairn really stared. A man was riding the debris, going past at what must have been 15 miles an hour, and then more people came by in the rising flow. Pitcairn rushed down from the tower. He held the train where it was.

All that afternoon and into the evening, passengers and train crew splashed around in the rain, trying to help the people passing by in the water, to little avail. They saved seven. More than one hundred went by. That whole time, Pitcairn would say later, he still wasn't giving the South Fork dam any thought. He figured only that the Stony Creek must be astonishingly flooded.

Water kept rising toward the track, and Pitcairn wanted to get two freight trains at Bolivar backed into a siding, but now even that move looked risky. He prepared the passengers to run for the high ground on the hillside.

But then, a little before 6:00 P.M., the water's current seemed to slow. Pitcairn had seen nothing of what was going on, just four miles up the line: he didn't yet know about the monster, what it had done, what was still unfolding. Even as fire burned there, and the great cauldron seethed, and the houses kept falling, and the long night of horrors began, Pitcairn was seeing, without knowing it, a monster begin to die here in the plain to the west of town.

Pitcairn did know one thing. Something very bad was going on in Johnstown.

A standstill in the water, then a slight lowering of the level, and at about 6:00 P.M. Pitcairn decided to head No. 18 back to Pittsburgh. But thanks to the water still on the track, it took a long time to get only as far as New Florence. There he stopped the train again.

Pitcairn had decided he needed to hear word, directly from Johnstown, of what was going on there, so he could send some clear word to Pittsburgh. If things up the valley were as bad as he feared, direct word might be hard to come by for some time. And at the rate the train was going, he might not get back home for hours.

By now he'd heard some talk among the passengers and the few people they'd saved that the South Fork dam must have broken. Pitcairn didn't know whether that could be true. Nevertheless, he'd telegraphed the rumor to Pittsburgh earlier, when he'd first heard it. Now he needed to know something definite.

So he held the train at New Florence until a message finally did arrive there, at about 10:00 P.M. A railroad man had made it from Johnstown as far as Sang Hollow, not a town but a small extension of the railroad, with a station and telegraph office. The man had come on foot, walking the railroad tracks above the river, to bring the awful word. Johnstown was wiped out. Many were dead and dying. Debris was stacked at the stone-arch railroad bridge at least forty feet high—and on fire.

And this wasn't just some especially bad flooding. The South Fork Club's dam had in fact broken. The members' beautiful lake had gone down the valley.

So that's how the industrialists of Pittsburgh who were members of the South Fork Club and the public at large, first locally but then thanks to the telegraph all over the nation, got the story. Still in his plush car at New Florence, Pitcairn composed a message and had it telegraphed to the Pittsburgh papers. Among other things, his message called for a public meeting in Pittsburgh the next day, to begin relief efforts. The suffering of the Johnstown survivors, Pitcairn noted, was bound to be extreme. Supplies and food and medicine and cleanup tools must get to them as soon as the railroad could get a clear track.

Yet even after sending that message, Pitcairn held the train in New Florence. While his latest message would get on the wire before midnight, and would soon make newspapers nationally, his earlier, less certain message had already started another kind of flood, that of reporters out of Pittsburgh coming eastward toward the scene of destruction. Just after 7:00 P.M., a train chartered by the *Pittsburgh Dispatch* and the *Pittsburgh Times* had pulled out of Union Station, sending reporters through the rain; another train followed, with reporters from the *Post*, the *Commercial Gazette*, and the *Chronicle-Telegraph*.

Of course, thanks to the flooding, they couldn't get very far. Stopped in Bolivar, the reporters detrained and started getting stories from the many people gathered along the riverside and in the station. Crowds had come out in the darkness, bringing gas lamps to marvel at a downed bridge, and at the wreckage and bodies and living people coming down from Johnstown, and to make what rescues they could. Even as the people up in Johnstown were undergoing that awful night of the flood, their story was already being constructed by the first reporters to get to Bolivar and start interviewing an excited, horrified crowd.

These newsmen were acutely aware, however, that they were still twenty miles from their real story. They decided to go on foot and wagon to the next big station, New Florence: that would get them at least closer to Johnstown. They could get stories there and use the station telegraph office to file everything they had, and maybe tomorrow they'd get into Johnstown itself.

About 3:00 A.M. these cold, wet, professionally dogged Pittsburgh reporters got to New Florence, and again they found a crowd at the railroad station, where No. 18 still sat, a crowd of passengers and locals with the wildest stories to tell, many of them true. Then, at about 4:00 in the morning, an actual couple from Johnstown appeared at the New Florence station, the McCartneys, and the reporters mobbed them. The McCartneys had been walking for almost twelve hours, having escaped the city right after the wave hit. They told the reporters that hardly a building was standing in the city.

That was the scoop. The newspapermen all ran to file.

But they'd also discovered that right here at New Florence was the head of the Pittsburgh section of the railroad, Mr. Robert Pitcairn himself, one of the most authoritative men in the region. They interviewed him, and Pitcairn gave the reporters, among other things, the name of the sporting club located above Johnstown, where, as direct word now made clear, a dam had broken, and he gave them the story of that dam, based on his own personal knowledge.

Under his own instructions, Pitcairn told them, the railroad's expert civil engineers themselves had been inspecting that dam every month, and all indications had been that the South Fork dam was as strong as any dam could ever be. Only an extraordinary natural disaster like today's record-high flooding could have taken it down.

Having given the reporters that news, Pitcairn ordered the train to move. A little after 4:00 A.M., it pulled out for Pittsburgh, going slowly to cope with water still over the tracks in the low grounds. Pitcairn didn't arrive in Pittsburgh until about 6:00 A.M. His interpretation of the cause of the flood had meanwhile attached itself to the early stories of the Johnstown Flood.

"Investigations showed," one reporter informed his readership, "that nothing less than some convulsion of nature would tear the barrier away."

◆ ◆ ◆

In Pittsburgh the next afternoon, while survivors in Johnstown were moving through the horrible landscape of death that yesterday had been their city, and the fires still burned at the bridge, a big meeting was held at the Old City Hall. The subject was the relief of Johnstown.

Robert Pitcairn addressed the people assembled in the hall and made an impassioned plea for urgent action. A Pittsburgh Relief Committee formed itself at that meeting, and in less than an hour almost $50,000 was raised in the hall, an astonishing sum, about $1.2 million in 2018 terms. Four members of the South Fork Club—Henry Frick, Henry Phipps Jr., Reuben Miller, and S. S. Marvin—were appointed to the relief committee's executive board, and it was Miller and Marvin, less august than Frick and Phipps, who would carry the ball, working hard to organize relief efforts.

And another meeting, a private one, was held that night. Some of the key South Fork Club members met at one of their palatial homes. The issue at this meeting was their dam.

Where might fault lie, for a degree of destruction whose scope wasn't by any means known yet? The specter of potential lawsuits, at the very least damage to reputation: none of that would have eluded any of these men. Club policy, arrived at in this private meeting, was therefore to contribute as much as possible to the relief effort and to say as little as possible about the South Fork dam.

What was going on in Johnstown that same day was just the beginning of what would become a ceaseless commitment to an extraordinary variety of horrifying, laborious, daunting, often even overwhelming work. Facing the horror on Saturday morning, it would have been hard to imagine gaining any improvement at all. Yet almost instantly, people started trying to better a disastrous situation. In that process, which would go on against terrible odds for weeks and months to come, the communities of the Conemaugh Valley came together in an effort to revive life in the face of death.

And yet frictions in those communities would also be exposed by the work that lay so dauntingly ahead. Nobody could think about it now, but what the bosses' club had so blithely done to Johnstown, and differences in how Johnstowners tried to cope, and even differences over relief efforts from outside, would cause unsettling change, even as people began working together to overcome a degree of adversity they'd never seen before.

The Reverend David Beale took a leading role in getting things going. Standing on the hillside, first thing on Saturday, with thousands of cold, homeless, hungry, and bereaved people, he'd sent the boy on horseback to telegraph the governor of Pennsylvania. That was just one of many beginnings of a collective effort on the stunned survivors' part to save lives, clean up, identify and bury bodies, and, in the end, to restructure their government and rebuild their city.

Another important leader in the recovery effort, with an approach very different from that of Rev. Beale, didn't live in Johnstown and hadn't gone through the flood. This was Tom L. Johnson, the dedicated monopolist who had begun to fear that Henry George's critique of capitalism got things more or less right about the way monopoly created poverty. Johnson's skeptical mind had long been worried about the South Fork dam, despite the reassurances of the club members, and he was in his house on Millionaires' Row in Cleveland when late on Friday night he heard the first news reports of the destruction of Johnstown.

With his partner Arthur Moxham, Johnson had built an important part of his business empire in Johnstown. Having become so adept at using special connections in government— this was how, as he'd learned, business empires were built everywhere—to benefit from special highway grants, tariff, patents, and monopolies on land, Johnson and Moxham had just completed building their new steel mill, via relationships with members of state and local government, as well as with the Cambria company. It stood southward and uphill from the city proper at that new town they'd so modestly named Moxham. They'd even bought all the buildable land there, with holdings reaching all the way to the nearly perpendicular hillside above the valley, placing the new mill where they might profit further by increasing land values that would come with the mill's operation and housing. That mill was just ready to start smelting and rolling on May 31, 1889.

On his hearing the awful news, Tom's first need was to find out whether his friend and partner was alive and well. He headed for the train station, and arriving in Johnstown on that horrible Saturday, as soon as train service had been restored, he discovered to his great relief that Moxham was unhurt. The next thing was to determine what kind of losses their company had sustained, and almost as immediately, what he could do to begin to solve problems, relieve suffering, and get the city back on its feet.

Johnson and Moxham's first Johnstown rolling mill had indeed stood right in the flood path, and their buildings had been swept away with everything else. But those buildings were empty when the wave hit: the machinery had all been taken up to the new mill. So they'd lost little in the flood.

Now Johnson and Moxham got to work. In the process of trying to bring Johnstown back, Tom turned from a monopolist into what some might call a socialist but what, to Tom himself, was just common sense.

At 3:00 P.M. that same Saturday, not only Tom L. Johnson, Arthur Moxham, the Reverend Beale, Beale's friend Captain Hart, and other authoritative survivors, but also a large mass of ordinary survivors of the disaster packed into a surviving building, the Fourth Ward School at Main and Adams streets, to try to determine a course of action.

The big organizational problem, to Tom Johnson, was the very phenomenon that had helped make his own fortune. City government here, as elsewhere, was fragmented. Each of the boroughs of Johnstown operated as its own little fiefdom, with its own councilmen, burgesses, and executives; they argued, both among themselves and with the other sets of independent officers, over personal jealousies and contested rights. With no overarching leadership to make the kinds of public decisions that might best serve the most people, the system, Johnson thought, worked inefficiently enough even in normal times. But in the face of such destruction, the system would be simply hopeless.

Temporarily, there had to be what Johnson had no problem calling a dictator: one man—both he and everyone else in a position to make public decisions took "man" for granted—with all legislative and executive power.

Others at the meeting thought so, too. In the annals kept by the people who now took charge, everybody agreed, but that probably meant everybody who, to them, counted. The leaders who emerged were all white men, certainly, and most of them were Protestants: such were the hierarchies of the day. The extent to which the small black population and big immigrant population of Greater Johnstown—the Poles and Hungarians and Bohemians and Swedes and others, or the Catholics and the Jews—were involved in public decision-making was small in normal times. In this crisis, the usual powers took the lead, and others either kept their objections to themselves or were ignored.

And yet Tom Johnson, again showing his incipient radicalism, took a dim view of this seemingly natural tendency to turn for leadership to so-called leading citizens. When, for example, the mass meeting voted to dispense with their governments and elect the dictator wielding all legislative and executive power, the choice was obvious: John Fulton, now general manager of the Cambria Iron Company, which was, in a sense, identical with Johnstown. But it turned out that Fulton was out of town. Tom Johnson reflected that in a crisis, so-called leading citizens are often the most worthless of the citizens.

In Fulton's absence, both Johnson, fast becoming a kind of maverick, and Rev. Beale, a more traditional sort, were pleased that the dictator elected was Johnson's partner Arthur Moxham. Like Johnson, Moxham was becoming eager for what they were both starting to think of as good government: civic decision-making with a goal of benefiting all the people. That Moxham was, for one thing, a British subject never crossed anybody's mind.

So Moxham took over the meeting and created six committees: Finance, Supplies, Morgues (with Rev. Beale one of its heads), Removal (with Johnson a head), Police (with Captain Hart a head), and Hospitals. They all got to work right away recruiting labor.

For the Reverend David Beale, the next ten days passed in a blur. Normally a fastidious man, all that time he wore only the clothes he'd worn through the flood. For the first thirty hours or so, he didn't eat. The constant labor and intense anxiety he felt may have helped him stand it, he thought.

His job was morgues, which meant finding spaces where bodies taken from the wreckage could be laid out, identified as well and quickly as possible, and buried as quickly as possible, too. Burial was urgent. In fact, the cool weather and overcast skies and rain—it had begun again—were delaying the decomposition of the corpses, but it was spring, and the sun might come out and heat things up at any time. The spread of infectious disease was yet another of the terrors that the surviving Johnstowners faced.

Tom Johnson, meanwhile, was in charge not of morgues but of actually getting the corpses removed from the pile. He found himself taking up that work with a gut feeling of horror, really a deep unwillingness. As Tom began directing his crews, and as he and they pulled the first few bodies out of the pile, he looked at the dead and started crying.

Then something strange happened. Very quickly, that first day, he lost any fear, any reluctance, even any sorrow. If anything, Tom knew, he appeared totally unconcerned, heartless. He and his crews just kept at the work.

As Johnson and Beale and many hundreds of others kept at their many awful tasks, around and about the vast pile of wreckage that Johnstown had become, more people started coming into town and more tasks were assigned. A relief committee train left Pittsburgh at 4:00 P.M. Saturday, nearly twenty cars packed with volunteers. It was slow going—all the trains that had been stopped for the flood were being moved out, too—and they didn't get to Sang Hollow till 10:00, and there they found the roadbed washed out.

Emptying the train, the volunteers came the rest of the way on foot, loaded down with food. They arrived on the west side of the Conemaugh early Sunday morning, and by 7:00 the train was there too, Robert Pitcairn having made it a priority to get the track from Sang Hollow rebuilt.

But everything had to stop on the west side of the river, across from town, near the big pile at the damaged railroad bridge—the pile was still on fire, and there was no going over the bridge anyway—and that became the first deposit point for food and supplies, the hub from which stuff could travel across the river, first by boat, soon by a precariously swinging rope bridge, then by a comparatively reliable pontoon bridge.

Meanwhile, the Baltimore & Ohio Railroad had secured an entrance to Johnstown itself, via the south side. Members of the Pittsburgh Relief Committee, now in Johnstown, had a telegraph working, and by wire back to Pittsburgh they ordered that all supplies for Johnstown proper be sent around by the Baltimore & Ohio, which was dedicating all of its local facilities to relief.

And soon between 6,000 and 7,000 laborers would be on the scene, well beyond the volunteers, men hired and brought in by the Pittsburgh committee and managed by contractors. About 30,000 Johnstowners, volunteers, and paid workers were being fed at a commissary set up to cook and serve massive meals, with the food also hauled in continuously. The contractors' men came with hundreds of wagons, and many of the wagons carried dynamite. Explosions quickly became constant, jarring the fragile nerves of survivors and workers alike. Some complained, but the contractors insisted that to break up the pile, only blasting would do. At the still-burning pileup at the stone bridge, fire engines sent from Philadelphia shot gallons of water. After some days, the fire would go out, but nobody could tell for sure if the fire engines had beaten it down or if it had finally run out of fuel and died of its own accord.

As blasting now shook the place that had been a city, and bodies were laid out in surviving churches and then carted off for quick burial, and trainloads of supplies were unloaded and deployed by thousands of workers, back in Pittsburgh the Ladies' Committee was accepting and finding housing for huge numbers of orphaned children and homeless and bereft people of every kind. The committee set up shop at the city's big Second Presbyterian Church and stayed open twenty-four hours a day to accommodate the train arrivals rendered unpredictable thanks to the ongoing effects of flooding on the tracks. Subcommittees met every train on both the Pennsylvania and the Baltimore & Ohio lines and carried people to the church, where other subcommittees fed them, gave them donated clothing, and found places for them to live. People leaving Pittsburgh to stay with distant family or friends got free transportation on the railroads.

The Ladies' Committee also handled that all-important thing, information. Some panicked people were joyfully reunited. Others had their worst fears confirmed.

If the first relief train to get from Pittsburgh to Johnstown didn't bring such great men as Carnegie, Frick, and Mellon, it did bring the kind of hands-on man that Tom L. Johnson, at any rate, couldn't have been happier to see: Bill Jones. Long the top Bessemer expert at the Cambria Works, before being hired away by Carnegie, Captain Bill was well liked in town and had brought a team of three hundred men from Pittsburgh, much equipment for setting up camp, and a host of tools. He was paying for this operation himself.

The other thing Jones brought was expertise in solving

problems and the kind of can-do energy that was harder for the bereaved to sustain. He and his men rolled up their sleeves and went right to work. Jones was a technical man, and he could see right away what needed to be done first and what could wait. And so Bill brought the kind of relief, Johnson thought, that money couldn't buy. Jones wouldn't get home for weeks.

Tom Johnson had heard some men in the monopolist community say that at the finer London hotels, where Bill Jones stayed when giving his lectures on steel innovation, he didn't cut much of a figure. Where Bill's boss Carnegie hung around in London with the philosopher Herbert Spencer and the poet and critic Matthew Arnold and had intimate chats with Prime Minister Gladstone, Bill the rough-and-ready steel master gave gritty technical demonstrations to other steel people on the chemistry and machinery involved in making steel. Bill was just the opposite, Tom thought, of the "leading citizen" type, so useless, regardless of how much money they might cough up in an emergency.

In his mind, Tom opposed the story of Bill Jones getting looked down on by capitalists and poets and politicians in London, with the sight he had of Bill now, working without pause to bring his old home back from horror. As Tom and Bill, dirty and sweaty from their work, were riding around the pile on horseback, a man hailed Bill, calling him heartily by name, and Jones dismounted, although the man, straight from working on the pile, was so covered with dirt that his face was unrecognizable.

"You'll have to tell me who you are," Bill said.

"I'm Pat Lavell," the man said, and Pat and Bill, overjoyed to see one another, fell into a hug.

Then the painful pause. Everybody in Johnstown had to endure it now, when reuniting with an old friend.

"How did it go with you?" Bill finally asked.

The Gilded Age humor magazine *Puck* takes potshots at millionaires living it up at the South Fork Fishing and Hunting Club while destroying America. CREDIT: *Puck*, June 19, 1889

ABOVE LEFT: Henry Clay Frick, steel and coal baron, founding member of the club. CREDIT: Library of Congress

ABOVE RIGHT: Daniel Morrell, the Johnstown ironmaster who warned the club about the South Fork dam. CREDIT: Library of Congress

ABOVE LEFT: Steel magnate Andrew Carnegie helped pioneer both American monopoly and a fad for recreation amid natural beauty. CREDIT: Library of Congress

ABOVE RIGHT: Tom L. Johnson. The Johnstown Flood changed him from a monopolist to a progressive. CREDIT: New York Public Library

Lake Conemaugh: 20 million tons behind a soaring dam. CREDIT: Johnstown Area Heritage Association

South Fork Fishing and Hunting Club members' lakeside "cottages." CREDIT: Johnstown Area Heritage Association

High-end recreation and relaxation, 1880s-style. CREDIT: Johnstown Area Heritage Association

Boating: a luxurious novelty in the Pennsylvania mountains. CREDIT: Johnstown Area Heritage Association

"Cambria-land": Johnstown's industry in the valley of the Conemaugh. CREDIT: Library of Congress

Harper's Weekly gives its readers an inside glimpse of the steel industry that was remaking America. CREDIT: Library of Congress

Men labor in the mills and mines, women at home. CREDIT: Pennsylvania State Archives

Immigrant coal miners in western Pennsylvania.

White tents of the relief camp can be seen in this wide shot of a ruined city.

A work crew poses for one of the many photographers who rushed in to capture the devastation. CREDIT: Library of Congress

The grim task of recovery. CREDIT: Library of Congress

Red Cross Hotel No. 3, Johnstown, Pennsylvania. CREDIT: Library of Congress

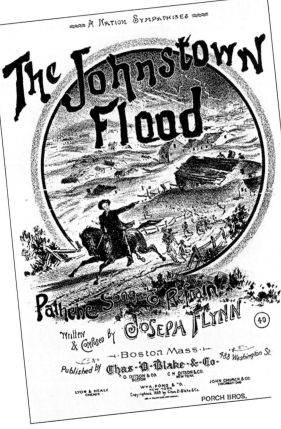

The mythical "Paul Revere of Johnstown" gallops across some sheet music. The scale of the disaster shocked the nation, and the flood soon featured prominently in popular culture. CREDIT: Johnstown Area Heritage Association

Publishers cranked out popular accounts of the disaster.

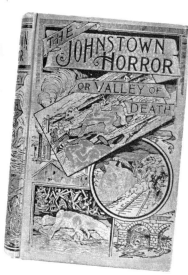

The public's hunger for special-effects disaster spectacle didn't start in Hollywood.

. . . but by 1926 the Johnstown Flood was on the big screen. CREDIT: Johnstown Area Heritage Association

"Lost everything," Lavell said. "My home, my savings, everything. But I'm the happiest man in Johnstown"—and here the darkened face lit up—"for my family's all right!"

To Tom Johnson, Bill Jones's encounter with his old friend Pat Lavell was worth incalculably more than any of the hotels where Bill was so looked down upon, and more than any of the people, for all of their wealth and art and literature, who did the looking down.

CHAPTER 12

POOR, LONE WOMAN

THE MAN IN CHARGE OF ALL OF THIS RECOVERY ACTIVITY IN
Johnstown was one Daniel Hastings, adjutant general of Pennsylvania. With no real military experience, Hastings was nevertheless a big, crisp, authoritative, and confident figure, with a
bearing befitting his title. Admitted to the bar in 1875, he had
a thriving law practice, and he was invested in coal mines and
banks. His only real command of troops had come twelve years
earlier in helping put down the railroad strike.

Hastings had arrived in Johnstown—he lived in Bellefonte,
seventy miles away—on Sunday morning. That day, it wasn't
clear how official his presence really was, but despite the election
of Dictator Moxham, Hastings was the state's ranking military
officer, and that morning he began a process—appreciated by
some, criticized by others—for taking semi-official charge of
everything. His first impression of the situation at the disaster
site was that armed forces were badly needed here. He collabo-
rated, at first, with Dictator Moxham, and proposed asking the
governor to send troops.

Moxham objected. He thought it was important for the people of Johnstown to manage crime problems themselves, through their own civil efforts, without an influx of outside police and military. That would strengthen the beaten people's self-reliance and maintain their accustomed democratically accountable local policing.

Rev. Beale disagreed. His very first action had been to send a message to the governor to ask for troops, and as his mad rush of work went on in those first few days, the reverend's main concern was law and order. As he went about the awful work of claiming standing buildings to use as morgues, laying out bodies, and comforting grieving families, Beale identified three types of crime.

One was simple theft of stuff lying in the wreckage, everything from clothing to silverware, from money to fabric, from coal oil to whiskey. Another was trophy-seeking: creepily enough, people were already coming to town from far away to get their hands on disaster souvenirs, small things like spoons, jewelry, hymnals, anything you could easily carry off that had actually been through the flood.

The worst, in a way, Beale thought, were the con men. People were already showing up in Johnstown offering distraught and homeless people, especially good-looking young women, all kinds of phony hopes: new homes, for one thing, if the women would only come away with them.

All of this had to be stopped, and to Beale, the need for a military level of policing was self-evident. The chief of police had been overwhelmed by grief at the loss of his family in the flood. Captain Hart, as a head of the police committee, was trying, and within eighteen hours after the flood, three hundred policemen were guarding two banks whose vaults reportedly still held more than $400,000. When the first trains began getting in from Pittsburgh, some police from that city were among the arrivals. But it wasn't anywhere near enough, Beale was sure.

Along with real crimes, rumors had started flying of invented ones, badly frightening the city's shaky people. Reports of atrocities were atrociously exaggerated: no fingers were actually being cut off by "human ghouls." Still, at a meeting on Sunday night, with Moxham arguing for his local police concept, men like the Reverend Beale supported General Hastings in his effort to get state troops onto the scene. Rumor had it that the looting was being carried out by "the Hungarians," a term that stood for any immigrant of Slavic descent. Beale knew that wasn't true—the thieving he'd witnessed involved all ethnicities—but it was also being said that in response to this demonizing of the recent immigrant population, vigilante posses were forming to kill recent immigrants who struck them as suspicious. Charlie Dick, the man who had insisted on getting his family up on the hillside, fallen asleep, and then witnessed the horror of the fire at the bridge, had been elected, like Moxham, "generalissimo" of a borough now cut off from Johnstown proper by the bridge failure. People were saying that Dick had taken to carrying a rifle and shooting thieves from horseback. That didn't sound like Charlie Dick, and Beale later ascertained that the story was nothing but a fabrication.

And yet it was by no means impossible to imagine, given the fact of real, ongoing theft, and drunkenness, and the general depression and displacement of all of the people, that English-speakers would turn violently against the immigrants: "fearsome Huns" and "worthless Poles" were other terms being used for looters. The community was coming together, as communities do, in the face of disaster. But the community was also coming apart, over anger and bias and fear that had long preceded the disaster and now, like the mud below the former city, lay exposed.

Then there was the sudden, ongoing influx of thousands of people from outside of town—many to help; plenty to steal, gawk, or run cons. The Reverend Beale wasn't alone in thinking that General Hastings had the only realistic solution to all of this rising conflict: bring in police and troops and regain and maintain order.

So the general overrode the elected dictator Arthur Moxham and sent a message to the governor, asking for troops. A few days later, Moxham stepped down and was replaced by a member of the Pittsburgh committee, James Scott, who added some new committees: one of them, the Department of Public Safety, was now to be run by General Hastings. Meanwhile, policing efforts started swelling in a chaotic manner. Pittsburgh sent trains with real police, but some members of the Pittsburgh Relief Committee visiting Johnstown put on stars they'd cut out of tin and appointed themselves police, too, walking around with baseball bats. It was hard to tell the real police from the self-appointed kind, and it wasn't clear who really had arrest powers. Shattered locals, exhausted by ceaseless work and overwhelmed by loss, were on the one hand grateful for the support of Pittsburgh, on the other hand at once frightened and irritated by the presence of more than one thousand varying, temporary, semi-official police, arresting journalists and assigning them to work crews and launching, on their own hook, a program of shooting all stray dogs and cats.

The surviving Johnstowners felt even further put upon by what seemed to be a prevailing idea among outsiders trying to help: that in the face of disaster, they'd suddenly become a lawless people, incapable of running their own affairs. The fact was that most of the crime—certainly the overwhelming number of the arrests—involved people not from Johnstown but from outside. Ordinary citizens struggling with unimaginable losses and challenges didn't necessarily enjoy all this friendly control imposed on them from Pittsburgh.

Soon the Pittsburgh relief man James Scott was out, too: the state had officially taken over now, and on June 12 General Hastings was put in official charge of the whole thing. In his hands lay all logistics of cleanup, relief, and law enforcement. By then, in the stark mud field that had once been the city's rectangular Central Park, Hastings had established a military camp for about six hundred troops of the state's Fourteenth Infantry Regiment, as sent by the governor at his request. By June 20, meanwhile, civic policing was back under the control of the city's police chief. That was a relief to many Johnstowners: the more familiar force—people knew their policemen—had been reestablished as well and was making its difficult rounds through the wreckage.

One of the biggest jobs requiring military degrees of planning, administration, and deployment was disinfecting the entire disaster area. Any house not swept away was filled, to at least the second floor, with deadly muck: whatever the water touched, it polluted, so the slime on the wreckage and the mud below made the whole scene a breeding ground for disease, thus also for dread. Remedying filth on such a vast scale couldn't, in the end, be a bottom-up or semi-voluntary project.

Huge supplies of many different kinds of disinfectant came in by train, not only from Pittsburgh but also from Baltimore and, especially important, at the order of President Harrison and Surgeon General John B. Hamilton, from the nation's capital. Dr. Hamilton also visited Johnstown and advised on medical care, disease prevention, and the uses of the different kinds of disinfectant.

A sanitary corps was established. It divided the city in twelve districts, each of which was assigned an inspector required to make a daily round and ensure that the disinfecting crews, under their foremen, were actually disinfecting. The district inspector was also to distribute disinfectants and instruct people in standing or only partially demolished houses how to apply them, in cellars, yards, and outhouses. Each district had its own office and a warehouse for storing the products, where people were urged, by placards posted all over the ruined city, to come and get the ones for domestic use.

Throughout all of the districts, the amounts and kinds of disinfectant sent, received, stored, and used were mind-boggling: thousands of barrels and bottles of quicklime, chloride of lime, bromine, copperas, carbolic acid, muriatic acid, nitric acid, rosin, barrels of pine tar, pitch, sodium hypochlorite, corrosive sublimate, phenyle. Also proprietary products, with brand names familiar in the day: Sanitas, Bromo-Chloralum, Phenique, Utopia, Purity. The proprietary stuff handed out for home use came in small packages with directions. The big stuff, especially bromine, was used on big, mucky, foul surfaces and for wide street sprinkling, sometimes in high, white heaps.

The workers' camps also had to stay clean. Thousands of people were using latrines daily, and those latrines had to be disinfected daily, along with the giant kitchens that fed both victims and workers. As the bodies were pulled and embalmed, and the wreckage moved to expose new layers of filth, and people scattered the products in the cellars and yards, the whole scene came to lie under blankets and piles of acrid chemicals.

Unlike the Reverend Beale, Tom L. Johnson was less concerned about the crime situation than about long-range planning for a different kind of civic future, one in which things like this disaster could not occur. As the former monopolist went about his grisly work of pulling out bodies, his natural ebullience somehow didn't fail him, and he was taking a different view from Beale, not only of policing but also of what a city really needed at a time like this.

For some time, Tom had been trying to get hold of Johnstown's streetcar franchise, just as he'd done in so many other cities. The current owners of the streetcar line hadn't wanted to sell. Now they did: with the city destroyed, the streetcar business was a liability, not an asset. Most of its cars and tracks were washed away. Tracks not destroyed were under debris. Early in the recovery, Tom Johnson not only bought the streetcar service but did so at a rock-bottom price. The sellers, he figured, thought he was an idiot.

But Johnson knew two things. One was that since the debris was being so laboriously cleaned up, by himself and others, sooner or later Johnstown would be back. In fact, he was finding it marvelous: utter destruction was already being ameliorated, and the short time that had passed since the flood suggested to him that complete reconstruction was not only possible but would take less time than anyone could have imagined. The ability of ordinary people to get things done was making a big impression on Tom.

So the value of the streetcar would have to rise. With an investment in new track, it would run again.

The other thing Tom knew: when it did run again, he would run it for free, at least for the first few months. He and Moxham had already made their private steam railway from the town of Moxham to Johnstown free to passengers, and Johnson wasn't noticing anybody around Johnstown talking about the sacred rights of property in the aftermath of the disaster. People were giving away food and clothing. Why shouldn't he give away transportation?

At first, as he developed this plan, Tom thought only that trying to make money on essential services when disaster conditions prevail is obviously a criminal act. He tweaked his partners by telling them the great thing about his free-of-charge approach: all fare collection problems endemic to the streetcar business would now simply evaporate. They replied that, on balance, they'd prefer to cope with fare beaters than lose income from passengers.

Then Johnson began to think that streetcars everywhere should be free of charge, always. The fact that a private enterprise could make money on the service had long been a source, as he knew quite well from successful experience, of capitalists corrupting government to gain the franchise. Nobody was trying, by contrast, to get hold of any fire department franchises: there were none, because there was no money in firefighting.

Maybe transportation was like firefighting. Maybe all essential services should be run that way. Funding to run the service could come from public money raised by taxes—possibly new taxes on land—and not by charging people who had no choice but to ride. For Tom L. Johnson, the horrible cleanup after the Johnstown Flood, which had been caused, he had no doubt, by the heedlessness engendered by special privilege, was turning into a laboratory for new civic thinking.

General Hastings had begun trying to run everything with military precision even before he'd been officially appointed by the governor to do so. And before his official appointment, he'd faced a weird disruption that he found hard to process.

On Tuesday afternoon, June 4, with troops sent by the

governor already arriving and being deployed throughout the wrecked town, Hastings found himself confronted by a small, slight woman, barely five feet tall, in a black dress. This poor, lone woman—that's how she struck him—announced to him that the Red Cross had arrived in the field.

The term "Red Cross" meant nothing to General Hastings. The woman's name—Clara Barton—might have rung a bell: she'd gained some renown over the years, but the Red Cross wasn't anything like a well-known or well-established organization. Barton had formed it herself, only eight years earlier. The Red Cross had performed some important relief work in Texas, but it wasn't the only organization of its kind, or the leading one, and while it had been chartered by the United States government, it had zero authority or reputation. General Hastings couldn't imagine that here in the devastation of Johnstown all that was about to change.

Clara Barton had found her calling during the Civil War. Not only did she raise money during that war to buy and transport nursing and medical supplies for wounded Union soldiers, she also somehow got personally close to the action. Women were never allowed there, but Barton performed hands-on nursing and deployed, under her own direct command, the distribution of supplies and nursing. She quickly became invaluable, and she gained official U.S. Army permission to operate just behind the deadliest front lines of the war at Antietam, Fredericksburg, and other scenes of vast, gruesome carnage. Not only did she actively tend, with her own hands, the most grotesque injuries of that war, but she briskly managed all logistics of relief.

After the war, she parlayed her legend into an organization with some funding. Traveling in Europe, she helped the Swiss Red Cross set up military hospitals during the Franco-Prussian War and began to see what an organization focused solely on medical care could achieve. She was decorated by Prussia for those efforts. Learning of Europe's Geneva Convention, protecting noncombatants and setting humane rules for warfare, she brought that treaty home and presented it to President James Garfield: the U.S. Senate ratified the Geneva Convention in 1882, during the presidency of Chester A. Arthur. Amid that success, Barton founded the organization first known as the U.S. Red Cross of the Geneva Convention.

That all took a long time. By the time she'd finally succeeded in her relentless effort to get the Red Cross up and running in the United States, she was sixty-one years old. And yet Miss Barton, as she was always known, remained tireless both in pursuing relief projects and in commanding her new organization. Her headquarters now was in Washington, D.C., and from there she'd led teams in bringing relief to famine and cyclone victims. She and her team knew how to put out an appeal, raise money, and organize delivery and distribution on a pretty big scale.

The Johnstown Flood would put those skills to the test on a whole new level. And it would put the Red Cross on the map.

The last thing Miss Barton could have known, when she first got the news about the disaster in Johnstown, on the morning of Sunday, June 2, was that it was about to change the national status of her organization. What she and her team had to do was verify whatever they could about the extent of the destruction, pack and plan as best they could, establish rough communication plans with their D.C. headquarters, and lay out a preliminary fundraising plan. That afternoon Miss Barton and five others splashed their way down to Washington's Union Station—the rain still hung on; much of the east coast was now inundated—and boarded a train bound for Johnstown.

They made slow progress. Track had been washed out in many places. They didn't arrive for forty-eight hours.

What they encountered on arrival was like nothing Miss Barton or anyone on her team had seen before. Each of the war zones and famine scenes and natural disasters in which she'd served had been uniquely awful. But Johnstown had been destroyed in the way that only later wars would make possible: nobody had ever seen an entire city laid waste all at once like this, nobody had ever seen the kind of wreckage that can be caused by 20 million tons of water dropped on top of a city full of people. The cleanup and feeding and clothing and burial had been under way for only hours at this point. The wreckage at the bridge still burned. This degree of misery was shocking even to Clara Barton.

She and her team began climbing their way over wreckage and wading through mud, on a quest to find General Hastings. They'd been told he was the person most in charge here. Miss Barton took everything in as they went, and while she would end up staying in Johnstown longer than she'd ever stayed at any other disaster site—it would be five months, and cold autumn, before she would see her home again—it was the shock of that first day she would never forget. In constant drizzle, she was climbing up broken locomotives, letting herself down over heaped-up iron-rolling machines, getting around bent rails, avoiding barbed wire. The team went past people carrying corpses out of the wreckage, animal carcasses. They could smell the acrid smoke, and the slime of the muddy water, and they knew the fear that must have struck the already terrorized survivors: disease. Miss Barton saw that all the businesses were destroyed or closed, the electric light out, most of the money in the bank vaults gone to the river bottoms.

And she saw that thousands had nowhere to sleep, nowhere

to get warm, not enough clothing. She continued on her way, intent on finding General Hastings.

Hastings was a military gentleman of the old school. Confronted by this strange little woman, obviously here in hopes of somehow helping, and speaking what sounded to him something like gibberish, his only thought was to do right by her.

Mentally the general now therefore added to his endless list of tasks some means of making this poor lone thing, who would be helpless in a place like this, as comfortable as possible. The general gallantly assured her that he would find every means of helping her and her friends.

Miss Barton did her best to clarify. This was the Red Cross. They didn't need to be helped. They were here to help, and they knew how to do so.

Hastings looked at her as if she were speaking another language. Nothing about Clara Barton fit anything he'd ever considered possible.

Instead of trying to convince him of anything, Miss Barton and her team got down to work. They'd been joined on the way by members of their Philadelphia chapter, bringing tents and cots so they could make camp. The Red Cross set up a headquarters of its own in the mud, using a crate for a desk, and began by receiving and distributing the trainloads of supplies their headquarters back in Washington had already begun raising and having sent in huge volumes. The raising of Red Cross volunteers had begun, too.

One huge tent served as the Red Cross warehouse. Over the ensuing days, the growing team took delivery of packages of food and clothing, went through them in the warehouse, and started a process of handing things out to people lining up to get them. The rain kept coming down: it was hard to keep things dry. The dynamiting was going on, yet for weeks a cart couldn't even pass in the street, so everything had to be humped and schlepped on the backs of strong men from the train station and in and out of the big tent warehouse.

Yet soon this seemingly impromptu crew known as the Red Cross, its team swelled to hundreds of volunteers, had in many ways superseded the Pittsburgh Relief Committee and the state commission in organizing the feeding, medical care, clothing, and shelter of the suffering. This organization seemingly self-created and self-appointed, part of no government or army or police force, was turning, during its five months in Johnstown, into the nationally recognized institution that it would soon become.

Something Miss Barton took on for the first time in Johnstown was the building of quick, temporary shelters. Her idea wasn't just to get shelters up but also to outfit each of them with cooking facilities, beds, chairs, and tables. The Red Cross took on some of the building, the commission took on the rest, but the Red Cross furnished all of the structures so people could live there. The commission put up three thousand simple structures. The Red Cross furnished those but also built "Red Cross hotels": six buildings one hundred feet long and fifty feet wide, actual hotels, with services, housing thousands of people at little or no cost to them.

All that took ordering, inventory, bookkeeping. Money was raised, of course, but big companies around the country donated in kind, too: beds, bedding, enameled cookware, chairs, all coming in by train. Newspapers raised money to buy cooking utensils in the thousands. The furnishings were taken delivery of at the station, hauled to the tents, and marked down in ledger books that also recorded the name of each family staying in each of the dwellings and a list of each thing they were given. Along with everything else the Red Cross did well under wildly adverse conditions came bookkeeping. General Hastings began to get what was going on.

Soon, to Tom L. Johnson's total dismay, there was too much money in Johnstown.

The relief effort, as organized by the Pittsburgh committee, the state commission, and the Red Cross, and aided by endless, ubiquitous press coverage from coast to coast—Johnstown was all that people everywhere talked about for months—became so successful that an amazing amount of stuff came into the city from big donors, ordinary people, church groups, companies, chambers of commerce, and every other imaginable source everywhere in America. Relief committees had been appointed in cities and large towns, and people everywhere seemed eager to contribute not only supplies, and not only money to buy supplies, but straight cash, adding up to an astonishing amount that was hard to keep perfect track of. The question soon facing the various committees was what to do with it.

Much of both supply and cash came through Pittsburgh: people around the country knew that the wealthy families there were organizing relief, and the Pittsburgh Chamber of Commerce took on the role of financial agent for managing cash donations. In Johnstown itself, the finance committee appointed at the Fourth Ward School on the day after the flood took charge of receiving that money from Pittsburgh and elsewhere. So as the electricity was restored, and the telegraph operated, and the dead were buried, and the newspaper came out, and ways were cut for passage through the pile, the finance committee faced this new, weird problem of how to handle all the money. The problem became another source of friction among the survivors, and between citizens of Johnstown as a whole and the outside entities that wanted to dictate to them the terms of their own relief.

It took only a portion of that money to pay for the monumental relief work that was being done. Much of that work was paid for elsewhere or provided voluntarily. Rumor had it, more or less correctly, that $3 million had been sent in. About $1 million got spent on relief. The money had been sent as charity, to go to people in need, and distribution of the remainder had to begin.

But Governor Beaver had started receiving questions from donors about how the money was being distributed. So he appointed yet another state entity, to be overseen by him: the Flood Relief Commission, in charge of controlling and distributing the money. This commission included not one citizen of Johnstown. The governor and this new commission sent word throughout the country that all cash contributions to Johnstown should now go exclusively to the commission, which soon determined that it would set up a system for paying about $250,000 of that cash to indemnify people, not only in Johnstown but also elsewhere in the region, for loss of property during the record-high spring flooding.

That idea was received with indignation in Johnstown. The money hadn't been sent to insure property owners against spring flood damage: such damage had occurred many times before and nobody had sent money. The money was charity, in response to the destruction of towns and the city by a lake that had poured down when the South Fork dam burst, leaving the city nearly nonexistent, homes gone, clothing gone, thousands dead, survivors bereaved. Using it to pay for general flood damage throughout the region was a total denial of what had happened, a denial of the deaths of so many, of the desperate condition of the people, and of the purpose of the money to help people made destitute by a horrifying disaster.

In response to all this complaint, Johnstown's own finance committee made an official request to the governor for the funds, and the state commission did authorize the committee to make a distribution of ten dollars per capita to the entire citizenry. The committee did so.

But a meeting was also held in Johnstown to address this untoward issue. The staid Rev. Beale and the maverick Tom L. Johnson were agreed: indemnifying people for losses is not the purpose of a charitable fund. People spoke against the governor's commission. One of the most insulting things was the commission's requiring a signed oath from the citizens detailing their losses, in answers to questions on a form, in order to get a distribution. This seemed inquisitorial.

Still, arriving at a good way to distribute all this money seemed elusive. If you give the most to people who can prove they've lost the most, you're rewarding the better-off at the expense of the poor, and the better-off might have other resources that you don't know about. If you give more to the poorer people, you have some responsibility to ascertain that they're as poor as they say they are, a prying and humiliating prospect. Tom Johnson, fed up with the whole money issue, took pleasure in startling the meeting by proposing that all the money be converted into silver dollars, loaded on wagons, then hauled out and dumped into the streets, where the people could scramble for it. That, he suggested, would have a result at least as fair as anything this long, boring discussion could arrive at. Johnson wanted the issue handled quickly—really, he wanted it to go away—because he felt this extra-money issue was distracting the citizenry from all of the other matters that were still pressing:

cleanup, rebuilding. The problem of too much money offended him. He wished people hadn't sent it.

In his later account, the Reverend Beale simply ignored Tom's oddball proposal. The meeting did go on, and in the end the committee created a less inquisitorial system—a number of the requirements were waived—for reimbursing the neediest, giving each person who applied a certain percentage of their estimated losses. The claimants were ranked by classes, with those considered, in the ethos of the day, the most vulnerable getting the most, and the least vulnerable getting nothing, in this order: those made widows by the flood with children; those made widows with no children; the aged, decrepit, and injured; those who had lost all their property who weren't widows; those who had sustained considerable loss; and young persons able to take care of themselves, along with persons who had a lot left. The distribution was made in two phases, a smaller chunk first, and then, with the classes even more granularly broken down, the balance of all the money.

Tom Johnson was just glad that the issue had finally been resolved. To him, this focus on money seemed way off the important point. All the money in the world wasn't going to bring back the dead or comfort the living. The event never should have happened in the first place. What kept getting lost, in dividing suffering by class and category, is that this thing had happened, and was still happening, not to classes and categories but to real people.

CHAPTER 13

FROZEN WITH FEAR

SOME OF THOSE PEOPLE WERE LUCKIER THAN OTHERS. AFTER Max McCachren made his mighty throw, tossing the little girl he'd found on the raft up to the man leaning out a window, his raft just kept going, rushing and turning right down toward the stone bridge where the fire was burning people alive. As Max neared that inferno, reflected and refracted by the choppy water, he could make out a place where the water, on the right side as he approached, had worn away the earth: at that hole the flood was pouring through the dam of debris like a rapids. Max leaned and paddled and kicked and he got his craft heading that way.

Arriving at the flume, he got the raft into it. He shot the rapids. Others were running that exit, too, clinging to floats, and many were thrown off, but Max slid past the entire fiery obstruction and was sent downstream past Johnstown. About four miles farther down, a crowd on the shore was working with poles and ropes to get people out. Max caught hold and they pulled his raft to shore.

And in the days and weeks following that strange and horrible night, as the work went on in Johnstown, and the reporters kept seeking out and publishing one amazing or distressing or inspiring story after another, Maxwell McCachren became famous, momentarily nationally, permanently in town. The story of the throw was gold.

Meanwhile, by asking questions and following up on word of mouth, James Quinn was able to find out who had rescued his daughter: Max, for one, but also the two men in the window who had carried her to the Metz house. They turned out to be Henry Koch, the proprietor of a hotel near the house where they'd rescued Gertrude, and George Skinner, a porter working for Koch. James was able to find and thank them all, and when the per capita relief payment was made, James split Gertrude's share three ways and gave it to those three men.

The press loved Gertrude's story. Seizing upon George Skinner, reporters ginned up the rescue into the tale of the innocent white Little Eva and the faithful black Uncle Tom from Harriet Beecher Stowe's *Uncle Tom's Cabin*. In 1889, that angle was good for jerking a few tears. Yet the angle had nothing to do with what had happened to any of the people involved, or with what they'd done. Stories like that couldn't get at the devastating emotional effects that the arrival of the gigantic wave, from the lake above, had on real people.

The real Gertrude, for one thing, wouldn't talk, for a long time, about what had happened between the time the water burst into the third-floor room, where she'd huddled in the wardrobe with Aunt Abbie and Libby, killed by the wave, and the time she was brought to the Metzes' house with the Bowser sisters and other flood refugees, where her father finally found her. When the subject came up, she felt frozen by fear.

On first being found by Papa and her sisters, of course she'd felt enormous happiness, relief, and comfort. That morning they'd gone together down Adams Street, the first street just above the flood level, picking their way through disgusting, slimy wreckage, with Gertrude at once exhausted by her experience and embarrassed by her mismatched shoes: she made Papa carry her most of the way to the Leises' house. There the family picked up the baby and went to the larger home of another clerk at the Quinn store, Mrs. Ludwig. At bedtime, Gertrude slept with Helen and Rosemary on a mattress on the parlor floor, while the baby, whose condition seemed not to have worsened at all, slept in an upstairs room. Tended by a grandmotherly German woman of the house, warm and cozy, Gertrude Quinn at last fell blissfully asleep.

But as the days and weeks went by, things weren't blissful. Despite their relief in finding Gertrude alive and unharmed, terrible loss and pain had fallen on the whole Quinn family, and Gertrude was the youngest of the Quinn children who were out of the infant stage: she felt it all.

The Quinns were well-off, and they were a bright, idiosyncratic, happy, and resilient family. They weren't, in fact, typical of the community, in the way that the papers of the day, always eager to define normalcy in terms of the well-off, might have wanted to present them. The vast majority of people in the Johnstown area with lives demolished by the great wave had far fewer resources than the Quinns. And yet the real effects of the lake that had been sent down the valley could be felt even in the lives of one of the more prosperous families in town.

Gertrude's mother, with the youngest Quinn children, had been staying with family all the way out in Scottsdale, Kansas. News of Johnstown came quickly, of course, to Kansas, yet news from James to Rosina couldn't go quickly: it took days to get the telegraph and postal service working. And what James Quinn had to tell his wife was much worse than the fact that Johnstown wouldn't be a place to live for a long time to come, or that the family store begun by her father was gone, along with all its inventory, cash, and assets, never to be recovered, or that the family had no home, literally no possessions but what they wore, and Rosemary's umbrella. Those losses would have been devastating enough.

But Rosina also had to be told, to begin with, of the injuries to her own mother: Gertrude's grandmother had been taken from the wreckage of her home with her face and head covered with blood. Her scalp had been taken off from between her eyes to the back of her neck. Thirty-two sutures closed the wound, and she would live.

But of course James had far worse to tell his wife than that. Gertrude, it was clear, had been the only person to escape from the Quinn house when it fell into the flood. Abbie and her infant son had been lost to the flood, and Libby had, too. Their bodies hadn't yet been found.

Worst of all: their own Vincent was gone, sixteen, brilliant, funny, enterprising, loving, their firstborn. He died because he broke away from his uncle's attempt at restraint—the uncle and cousins and others had narrowly escaped the collapse of that building—and tried to make his way through the water to help his father save the younger children. Only Vincent's straw hat had been seen after the wave hit.

About a week after the disaster, Vincent's body was found, not far from the corner where on May 31, in a deep and turbulent current, he'd last been seen. That was in the early days of the massive morgue effort: there was no time for a real funeral, and the corpse of the Quinns' son was carried in a butcher's wagon to the cemetery and quickly buried.

The Quinn parents had been communicating by now. It was decided that the two younger girls, Rosemary and Gertrude, should go west to their mother in Kansas. Gertrude could hardly wait to see Mama, and their uncle Edward soon arrived in the devastated town to escort them west. For the trip, the girls were taken down to one of the relief commissaries to get clothing—they had nothing of their own—and as they stood in line and were handed outfits, it was actually exciting: warm, clean flannel petticoats, stockings, and shawls. The girls sat right down on the street and put everything on.

James Quinn watched. His girls were in clean clothes. That was something he was now in no position to provide them.

Over the slimy wreckage and through the muddy puddles they went, smelling the new smells of putrefaction and the harsh smells of lime and disinfectant, to the station to board a passenger train, refugees from the Johnstown Flood. With Rosemary and Gertrude sitting side by side and Uncle Edward behind them, they began the long ride westward. As the train picked up passengers from the stations west of the Johnstown area, the girls became the celebrities of the train: flood sufferers, children at that. The people gathered around to pepper them with questions.

Rosemary asked Gertrude to tell what had happened to her that day and night. As usual, Gertrude wouldn't, so Rosemary launched into a narrative of the family's sufferings. The people listened closely, silently, grimly.

When Rosemary finished, the people started handing over money, pouring it into the girls' laps. There was so much that the girls had no way to hold it, and there was no false pride: the Quinn family had nothing now. Uncle Edward gratefully filled his pockets with the money to give to their mother for the family's support.

In the house in Scottsdale, Gertrude got a new shock. She'd only just happily reunited with Mama when it turned out that Mama was going back to Johnstown, to attend a belated funeral service for Vincent.

Gertrude exploded. She cried and wailed and refused to stay here with her aunt and uncle. Mama sat with her and held her and explained, but Gertrude insisted through her tears that wherever Mama went, she had to go, too.

But the next day she looked for Mama, and Mama was gone.

Now Gertrude could not be comforted. She cried all day, every day.

And she started looking for her mother. She left the house. She walked down the street. Found and brought home, she would leave again, time and again, to find Mama. Her aunt had five small children of her own, and now Rosemary and Gertrude, so the neighbors around the house began to recognize the little flood survivor, and they took part in watching out for her in her wanderings and trying to keep her within some kind of bounds.

Gertrude, as usual, wasn't easy to deter. Neighbors brought things to distract and cajole her: a big chocolate layer cake that she ate with her fists and turned into a mess, dolls, toys. There was a raccoon in a cage in a neighbor's yard: the girl could be distracted by its antics for about a minute. Then she'd be crying and running off to find her mother.

One day she heard a train whistle. That train might get her back to Mama and Papa. Gertrude bolted from the yard where she and her cousins had been playing, ran as fast as she could down to the nearby train tracks, and sat down on the rails to wait for the train. She could hear it coming.

Her cousin Stella, seeing that Gertrude was missing, ran outside and spotted both the girl sitting on the rail and the locomotive rushing up the track in the near distance. Stella ran down to the tracks, snatched up Gertrude, and backed up against a fence, out of the way, as the locomotive rushed past them, followed by a long, pounding train of freight cars.

Mama came back. She was different. Gertrude was used to her mother's smiles and tenderness, but now it was Mama who cried, often, hugging Gertrude and Rosemary and the other children.

Still, Gertrude felt happy in those hugs, protected.

And they were all going back—not to Johnstown yet, but first to Latrobe, Pennsylvania, to stay with a family there, and then to a series of welcoming homes in Pittsburgh. Neighbors came in to help make clothes for the family's journey, and soon they were back on the train, heading east.

When September began, the Quinns were together in Johnstown, and Gertrude Quinn began to take part in the strange life of loss and recovery that all of the children of the flood would grow up with. The return was shocking at first. They entered their city from the station, their nostrils assaulted by the piles of powerful white disinfectant, a word that Gertrude and other children her age would quickly come to know. They were walking on planks that bridged puddles, mud, empty lots, open cellars.

So much had been done. There were electricity, telegraph, very partial streetcar service (free of charge, thanks to Tom Johnson), ways through the wreckage. Miss Barton and the Red Cross were still in town: thousands of new structures had kitchens and furniture. There were the Red Cross hotels.

But nothing was recognizable to Gertrude. They were walking to a new home, and as they made their way Gertrude saw a team pull a body out of the wreck. It was a young woman. Gertrude looked. Pieces of red-and-white gingham dress stuck to the body. She was hurried on, and at Central Park they found the park gone, no trees, just a mud field covered with tents and patrolled by soldiers.

The new house where they arrived was a kind of stopgap. Or at least James Quinn hoped so. He'd spent all this time trying to find ways to regain some kind of stability. The Quinns and Geises were propertied people, and all their property was gone. Everything they'd owned had been in the center of the town: not one of those buildings stood. Still, they were lucky: James had scared up the money to buy a small, yellow frame house. Its previous owners, a bride and groom, had died in the flood. The Quinns moved in. School started, as best it could.

Gertrude found she'd lost friends. All the schoolchildren had. She'd had a friend also named Gertrude, and two other playmates, Edith and Irene. All three were gone.

The surviving children played, however. And the city they lived in offered strange wonders and attractions of its own now. The temporary housing was funny: there was one kind, very flimsy, called an Oklahoma house, with one room and an attic on top, and there was a round old lady living in one of them. Gertrude and the other kids liked to play outside the lady's house and offer to do errands for her, in order to watch her go wobbling up the ladder to the attic, barely making it.

There was treasure to be hunted, too. Some men burying the guts of a sheep they'd killed had found a crock holding $6,500 in gold. The kids discussed who might have owned that money, and they started looking for finds of their own. If not a fortune, at least some fascinating objects: every yard had something; all you had to do was dig. In a single day, Gertrude once dug up a batch of old coins, three pennies, some nickels, and jewelry, including a shirt stud with a small diamond. Years after the flood, her sister Marie—once a baby with measles—was playing baseball with a bunch of boys and tripped while running to a base. She looked back and saw something shiny in the ground. It turned out to be a gold piece worth $2.50.

So the Quinn family, battered and set back, stayed in Johnstown. Diminished, they survived.

Rosina Quinn had always missed being in business. Now she started a new store, with James, and they did well. Rosina gained a reputation all over town for her complete grasp of commerce and great toughness in the matter of trade. It was said that nobody who dealt with her ever bested her in a bargain.

Gertrude Quinn grew up as the new Johnstown grew up. The family had no desire to leave. This city was their home, and James told Gertrude the day he found her on the Metzes' porch, "We'll be together."

Victor Heiser, by contrast, had realized on the day after his amazing ride through the debris that at sixteen he was alone in the world. As those first days went by after the flood, he kept searching for his parents. But now he knew he was looking for their bodies, in hopes of giving them burial. He stayed for a time with his friend's family, but he wandered the Johnstown wreckage, day after lonely day. He looked at hundreds of corpses. His anxiety, tension, and misery were extreme, but this was his task, assigned by himself, and he undertook it without stopping, for two straight weeks.

After a week, his mother's body was found. The Heisers had a cemetery plot that had survived the flood, and while many of the dead of Johnstown couldn't be identified, and were buried in a graveyard known as "the plot of the unknown dead," Victor was able to have his mother buried in the family plot. The boy kept up the search for his father, but after about another week of restless wanderings through the pile, bleakly staring at the faces of the dead, he had to give up.

Sometime later, news came to Victor that his father's body had been found, but nobody would let him look at it: the corpse was a grotesque mess. So Victor would never know whether his father's body had been identified and buried or not. Searching the ruins of the family home he'd seen collapse when the wave hit, he found a wardrobe, and hung inside it he found his father's blue Civil War uniform. That, and a penny in a pocket of the uniform, were all he had from his parents.

There was nothing to keep Victor Heiser in what had been the Johnstown he knew, nothing to anchor a life in the process of the great changes the city would now begin to undergo. Within a year Victor Heiser would be on his way.

Later, people would write about the Quinns and Victor Heiser and the other characters whose stories became well-known aspects of the Johnstown Flood literature, in part because both Gertrude and Victor would grow up to be good storytellers. Gertrude at last did become able to tell her story; one day she would publish a book about it. Rev. David Beale too wrote a book. In old age, Victor Heiser would give the writer David McCullough a long, amazingly clear interview on his experiences in the flood.

But there were so many other survivors, less advantaged than either Gertrude or Victor: perhaps less adroit in relating what had happened to them, perhaps less appealing to reporters at the time or since. Their stories were lost, or were less thoroughly documented, or went entirely undocumented. Their lives too were shattered in the flood. Some would stay in town, and some would leave. The struggles were so many and so hard—with more than two thousand dead, and so much property lost, literally everybody in the region was directly and acutely affected—that the people of Johnstown and the Conemaugh Valley, as well as the whole fascinated nation, had good reason to think that the perpetrators of such an overwhelming disaster were sure to be held somehow accountable.

CHAPTER 14

STRICT LIABILITY

F ROM THE DAY OF THE FLOOD ITSELF, ALL THE WAY THROUGH THE amazing relief and cleanup effort, and well into the beginning of the rebuilding of Johnstown, the question raged. What had caused the flood?

And with that question came the even more unsettling question of who, if anyone other than nature itself, bore responsibility for the enormity of the suffering, visited in so many different ways, on so many different people, the living and the dead alike, those who got to tell their stories and those who never did.

The answer to both of those questions was glaringly obvious to people as different in their thinking on many other matters as James Quinn, the Reverend Beale, and Tom L. Johnson. To them, the cause of the flood was obviously the failure of the South Fork dam. And the cause of the failure of the South Fork dam was that the dam was defective. And the defect was caused by the club members' negligence. A large number of people both in Johnstown and around the country agreed.

Later hydrology and physics studies would prove their view

correct, far beyond any reasonable doubt. Yet many people at the time, even in the absence of such studies, were already fully aware that while an earthen dam, if only reasonably well constructed, may hold back an amazing pressure of water even at full flood stage, once the dam is topped, it will go, and all of the water will come out. The way to avoid that is by installing release valves. That was Dam Hydrology 101 in 1889.

Also, many people had seen the South Fork dam's obstructed spillway. Many had noted the removal of the waste pipes and the infill that Ruff had installed where those pipes had been. With no way to actively lower the water level, the dam was due to get topped sooner or later, and so was due to fall sooner or later. That fact had been called repeatedly to the club's attention, beginning long ago, by Daniel Morrell's engineer John Fulton, for one: "a question of time." An employee of the club, Herbert Webber, had come to Johnstown earlier just this spring and met with burgesses and reported that over the years he'd brought the faulty condition of the dam to the club's attention: finally they'd told him to stop or he'd be fired. After the flood, Webber reiterated that complaint. To many people looking at the question of accountability, the case looked open-and-shut.

The club's position, however, had first been floated by Robert Pitcairn on the very night of the flood. Holding the train in the rain at New Florence, Pitcairn had told the first reporters, before they'd even arrived at the site of the disaster, that he had personal knowledge of the dam, that it had been inspected at least monthly, and that no dam on earth could have withstood the onslaught of this never-before-seen degree of flooding. The destruction of Johnstown was a tragedy because nobody could have done anything about it.

That was to be the club's line, in a conflict that now erupted, exposing even bigger conflicts between the Pittsburgh millionaires and the people of western Pennsylvania, conflicts that would mark American politics, law, and culture for decades to come.

The court of public opinion made its ruling on day one. When night fell on Monday, June 3, with the survivors still in utter shock, a gang formed in Johnstown and climbed up the mountain to the club property. Nobody was there: the few club members who had been there during the flood had taken horse for Altoona in order to avoid going down to Johnstown. The gang broke into some cottages and destroyed furniture, then went to Colonel Unger's farmhouse overlooking the former lake, now a huge, low expanse of mud with the South Fork Creek still running down the middle. The house was empty. The men hung around outside awhile and then went away.

Unger wasn't there because only hours earlier, during the day on Monday, he'd actually come down to Johnstown, with John Parke. Having recovered from the prostrate, overwhelmed condition into which he'd fallen when seeing the dam break and the lake start to roar down the valley, Unger was on his way to Pittsburgh, Parke home to Philadelphia, and they were met at the Johnstown station by reporters. Unger told the press of the desperate efforts that he and the others had made at the dam all morning and into the afternoon. Parke said that nobody could be blamed for what was a great, unavoidable calamity.

Also that Monday, reporters had preceded the indignant gang up to the dam site. They inspected the collapsed earthen thing that had been a dam, took note of the empty lake, looked over the clubhouse and the cottages. They also went into the town of South Fork to get stories from people who knew the dam and the club, and from people who had been at the dam when it broke. At the South Fork telegraph office, the reporters put those stories on the wire.

That's when the story of the club and its dam started blowing up from coast to coast. Stories told by local people painted the club in a highly unflattering way. The fact that the club members had bolted when the victims in Johnstown were going through hell to repair damage that those clubmen had caused, and to bury bodies their negligence had killed, put things in an especially ugly light. The members had always been high-handed and rude, people told reporters, monopolizing the fishing and kicking out boys who snuck in to throw a few lines. Respect was not the tone in which the locals described the members of "the millionaires' club."

But the big story, soon to drive long headlines over front-page stories from coast to coast, had to do with the faulty way in which Ruff's team had rebuilt the dam and the club's refusal to listen to reason and repair it. People from the region told reporters about the removal of the waste pipes, the infill of hay, the lack of cogent engineering, the warnings given, the years of fear that this spring, or the next spring, might be the year the dam fell. And they told about the fishguard, whose sole intention was to keep the lake stocked with prize game by deliberately obstructing the spillway.

Editors wanting to run these stories as fact needed independent engineering corroboration. They got it, from engineers sent to inspect. The New York *Sun,* however, scooped many awaiting further confirmation with its June 5 story headlined "Cause of the Calamity . . . The Pittsburgh Fishing Club Chiefly Responsible." Meanwhile, a Cambria County coroner's jury had formed, conducted interviews, and heard testimony in court, and the story was the same. On June 6, the jury issued its verdict: "said owners of the dam . . . are responsible for the fearful loss of life and property. . . ."

That did it. Papers in every city exploded with headlines like "The Club Is Guilty," "Engineering Crime," and "Shall the Officers of the Fishing Club Answer for the Terrible Results . . ."

Many of the articles under these headlines did miss a few points. Some said, for example, that the dam hadn't been built high enough, or that the thing wasn't strong enough to withstand the amount of water it held back. That wasn't it, as the engineers' report had made clear. The real issue had to do with what John Fulton had predicted years before: with no way to let the water out, the dam was sure to be topped one day. Then it would go.

But the papers got the bottom line right: neglect.

The think pieces followed fast. The horror of the disaster, the worst ever seen in the United States, was making its painful losses and thrilling rescues front-page news anyway, but the story of industrialists' heedlessness, in their pursuit of fun, leisure, and the restoration of their souls, wiping out the lives of people who worked below in the very industries that made the millionaires' millions, was so genuinely outrageous, so truly stark, that outrage flowed freely from the pens and banged hard from the typewriter keys. Welling up from these articles was all the fury harbored by ordinary Americans at the fabulous lifestyles of the captains of industry who were changing the face of American business and the American landscape. The disaster in Johnstown exposed in lurid, grotesque fact what some had already, using metaphor, described: the pursuit of capital to the exclusion of all else lavished life in huge, beautiful dollops to a select few by grinding up the very lives of the many.

Now it was literally true. A mine cave-in or factory explosion was bad enough: death and disaster in the service of profit. But

Johnstown was death and disaster in the service of relaxation and enjoyment, the reward of a few at the highest possible cost to the many.

So the worst crime wasn't, in the end, the flagrant negligence involved in constructing the dam—"that Desolate Monument to the Selfishness of Man," as one writer put it. The worst crime, as the story developed, was the lake itself, as created by the dam. The millionaires' fishing, sailing, and cooling off in the summer—the lake, literally—had come down the valley, killed all those people, and plunged all the others into homelessness and orphanhood, widowhood and childlessness, human sorrow and misery. From sober editorials to angrily purple op-eds, from cartoons showing a caricatured millionaire swilling champagne while Johnstown drowns to sarcastic doggerel poetry, the national popular and intellectual culture expressed collective revulsion at the very existence of that institution, whose name many Americans now knew, and could repeat with disdain and loathing: the South Fork Fishing and Hunting Club.

In this light, the members of that exclusive club were anything but enviable. They prized their fish and their fishing more than the lives of thousands of fellow human beings. They were beyond the pale.

That wasn't a controversial reaction. General Hastings, anything but a firebrand anticapitalist, told a reporter, "It is a case of carelessness, I might say criminal negligence."

Editor Swank, putting out his *Johnstown Tribune* a few days after finally abandoning his live-coverage essay, less than an hour before the great wave arrived in town, was one of the first to clarify the difference between a natural and an unnatural disaster. He noted the narrowing of the streams by industrial waste and landfill; the stone railroad bridge, with its visually impressive arches forcing the river through too little space; and then, of course, the failure of the unnatural dam and the fall of the unnatural lake. Thanks to clubmen who wanted to while away the summer days, "in all their spotlessness and glory . . . our misery is the work of man," he wrote.

Tom L. Johnson, meanwhile, noted that Robert Pitcairn's assertion that the dam had been inspected every month by civil engineers really meant that the club knew the dam was faulty, and thought it needed constant checking. The independent engineers called in to inspect after the flood reported that Pitcairn's engineers lacked engineering competency. And the independent engineers didn't believe the dam had been inspected regularly anyway.

The club took a different view, of course. On that first Saturday night in Pittsburgh, when after the mass meeting to establish the relief fund, the club members had met privately to consider their plans, they'd established a tactic: contribute to relief, and don't talk publicly about the club, the lake, or the dam.

That policy was carried out in bumpy fashion. In the end, only thirty-five of sixty members would make contributions to Johnstown's relief. Many of those who did contribute didn't exactly amaze anyone with the size of their contributions. With Frick's pretty impressive $5,000, and Carnegie's mighty $10,000 plus funding the rebuilding of the library, the Mellon family gave only $1,000, and one member donated $15.

Frick, a founding member, declined all interviews. As his great-granddaughter would note years later, Frick's membership in what had become a nationally odious organization would never be mentioned in her family.

In Paris, the United States legation to France put out a statement on Johnstown. Andrew Carnegie wrote it. The statement expressed sympathy and praised the heroism of the people of Johnstown. Nobody even knew that Carnegie was a member of the club. He didn't bring it up.

But while Frick, Carnegie, and some other top club members played things close to the vest, the club policy of saying nothing about the dam was difficult to enforce. "I am a member of the South Fork Fishing Club," one James McGregor told the *Pittsburgh Dispatch*. And he actually seemed to insist that, regardless of appearances, the dam was still standing. "I am going up there to fish the latter part of this month," he explained.

McGregor also reviewed the safety situation at the club.

"We have all been shaking hands with ourselves for some years on being pretty clever businessmen," he pointed out. So the members of this club, he reminded the paper's startled readers, were hardly likely to drop money anywhere unsafe.

Others went even further. Two club members announced that it must have been some other dam that had broken. "All the engineers," one claimed vaguely, were in doubt about whether it was the South Fork dam that had broken. The other noted that he'd been all over that dam, many times, and was planning on continuing to doubt that it had let go.

Other members might have found it especially worrying that the latter of those comments, denying that the dam had broken at all, came from James Reed, a partner of the club's own law firm, Knox & Reed. Both lawyers were club members, and the club's official position was to acknowledge that the dam had broken—it wasn't actually deniable—yet had broken only because no dam wouldn't have broken in such flooding: there was nothing, that is, that anyone could have done to prevent the unleashing of the monster, so there was no liability, personal or organizational. Reed's comments were way out of line, and by the time lawsuits started heating up, the club's position would be subjected to some serious discipline.

Part of that discipline was on view when Colonel Unger spoke to the press again, on June 5 in Pittsburgh. This time he didn't review efforts to cut more spillway and build up the dam. Instead, he pointed out that the original builder of the dam wasn't the club but the state.

The point of introducing such confusion as to chain of cause and responsibility would be to make clear to anyone considering a lawsuit that a suit might be long and difficult—thus expensive—and in the end potentially futile. Similar attempts at preemptive discouragement came from the club's lawyer James Reed, who spoke to the press again on June 12. This time, Reed didn't repeat his former statement that the dam was still standing. Instead he pointed out that the club had no significant assets.

That was true. Reporters had looked into the organization's finances and discovered that the club held only $35,000, with a $20,000 mortgage on the clubhouse. The members had millions, but the club, by design, didn't, so suing the club would be a pointless move, Reed stated. He was speaking, he said, just as a lawyer: if the shoe were on the other foot, he'd advise anyone coming to him in hopes of suing the club to drop it. As a lawyer, he couldn't see any liability, but should there be any, the total amount couldn't exceed the club's small capital anyway.

Daunting stuff, for anyone seeking to hold the members accountable. Still, the conclusions arrived at by such disparate observers and victims as General Hastings, editor Swank, the Reverend Beale, Tom L. Johnson, the coroner's inquest, the independent engineers who inspected the dam after the flood, the people of Johnstown, and the nation as a whole remained firm. The club's negligence was so arrogant and cavalier, the flood so preventable, the devastation it caused so disproportionately monstrous, the effort to reclaim the city so unprecedentedly challenging, that when the lawsuits did begin, many people had reason to hope that the richest men in the United States would be forced to pay for the horror, grief, and misery that their negligence had caused.

That's not how it went. Nancy Little was the mother of eight children; her husband died in the flood. She brought suit in Pittsburgh against the South Fork Club for $50,000 in damages for negligence. The club's lawyers pled not guilty and argued that the flood was an act of God. The case went on for years, but in the end the court ruled for the club. That put the club in a good position for handling other suits. When a consortium of Johnstown business owners explored the possibility of suing the club for damages and loss of property, lawyers spent two years studying the situation and then advised the group that the club couldn't be beaten in court. The group got a new lawyer and raised more money, but three years later the case was dropped.

A lumber dealer named Jacob Strayer filed suit against the club, first for $80,000, then upping it to $200,000. Strayer retained a Pittsburgh lawyer who died while the case was in progress. After a few years, Strayer learned that his lawyer had settled the case for $500, which he pocketed as a fee without notifying Strayer. A Johnstowner named Leaky sued the club to no avail.

Later, there would be no way to reconstruct, in any detail, just what went on to make it impossible to collect damages for something as overwhelmingly awful as the 1889 flood, when the cause of the event, the club's blatant negligence, had been attested to by coroners' juries and independent engineering experts. Record keeping was poor. The progress of each case to its inevitable failure will remain forever obscure.

Of course, the members were among the most powerful people in the United States. They owned everything in Pittsburgh, and that's where the suits were filed. The club had no assets, and nobody was realistically about to sue Henry Frick or Robert Pitcairn or Andrew Mellon or Andrew Carnegie for damages—even if their membership in the club could have been proven—on the basis that they were personally responsible for a general club failure to repair the dam properly, hence personally responsible for the unspeakable fate of Johnstown. No lawyer would have attempted it.

And that was the problem. As the lawsuits failed or were dropped as the futile efforts they really were, a feeling developed around the country—and in courts considering precedent and how to apply the law—that the law might as well have been written to protect those responsible for the Johnstown disaster from ever having to accept responsibility for it. The feeling was that the broad body of law informally called "the fault doctrine" of the day had abjectly failed the Johnstown victims.

Legal scholars would argue about whether that feeling was based on an informed legal reading. And yet the feeling changed the law. The astonishing horror of the Johnstown Flood of 1889, and the stark fact that nobody responsible could ever be brought to account for having so obviously caused it, turned a big page in the history of liability law in the United States. And the shift in liability law shifted a lot of other things.

Cold comfort to those who suffered loss and misery, to the highest degree, thanks to the flood, and failed to collect anything in return or hold anyone accountable. Yet because the newspapers, in particular, brought a good, strong tone of outrage to playing up the idea it had been impossible for victims of the 1889 flood to collect damages through accepted legal processes regarding fault, definitions of fault shifted in the 1890s and into the twentieth century.

The prestigious *American Law Review* agreed with the papers. The fault doctrine of the day, it argued, had defeated justice. Courts responded. Not only in Pennsylvania but also in many other states, they suddenly began adopting an expansive reading of an English doctrine known as the Rylands rule. That rule had developed about twenty years earlier, as a court decision in response to deadly dam failures in England, which killed people in the low hundreds, nothing like Johnstown. Also, the English dams that had failed were serving industrial purposes, not fishing and sailing. Still, under circumstances where an owner of property makes major changes to the property, turning something natural into something unnatural, the Rylands ruling stated that new liabilities kick in if something then goes badly awry, regardless of the conscious intention of the owner. While the right to enjoy your property may seem absolute—it certainly did seem so to big men like Ruff and Frick and other members—if you change that property in certain extreme ways, affecting others to their detriment, that right may not be as absolute as you might think.

Some things, that is, can't be blamed on God—even when the weather's bad. Sometimes people do things to change the natural situation in ways that, regardless of intention, create human responsibility. Specifically citing the failed dam and the falling lake at the South Fork Fishing and Hunting Club, rulings in a multitude of new U.S. liability cases began treating the bizarre events of May 1889 as not necessarily bizarre at all, but actually emblematic of a big new problem: industrial-age hazard.

For there was, in that sense, something new in the world.

Awesome disaster by natural causes—earthquake, hurricane, volcano—had been known since the days of Pompeii and before. But the scale of disaster caused by human interaction with nature, on the industrial level, seemed different, and it was the 1889 Johnstown Flood that first brought that difference home. When a giant, human-created wave can lift locomotives and barbed-wire factories off the ground, rain them down on cities, and explode into flame with molten steel and natural gas, something new is going on. A government of the people has to respond.

So thanks to Johnstowners' failure to collect damages, and public outrage over the Pittsburgh industrialists' outrageous unaccountability, courts started applying the Rylands rule in gas explosions, bursting water tanks, exploding nitroglycerine, and floods caused by coal mining. As industrial monopolies boomed in the 1890s, even beyond anything they'd achieved before, the industrialists' liability boomed, too, especially regarding what courts called "the unnatural." States imposed strict liability on mining companies, gas companies, and railway companies.

Johnstown became a legal mantra. Judges on the Pennsylvania Supreme Court got sick of hearing plaintiffs' lawyers waxing eloquent about the horrors of the Johnstown Flood. Other courts were sick of it, too: from precedent to mantra to cliché, the Johnstown Flood quickly become legally trite. That's because the courts had really learned what the culture at large had determined right after the flood. The Johnstown Flood represented a new and particular kind of crime. It was a crime against Gertrude Quinn and Victor Heiser, and against their families, and against all of the other people, known and unknown, that the flood affected. And yet nobody had ever been found guilty. The impunity of the South Fork Club, far from serving as precedent for further impunity, caused reaction instead.

Thus the 1889 Johnstown Flood served as the prime trigger for new relationships between big industry and liability, and between big industry and the people of the United States. For too long, courts, the press, and American institutions generally had excused industrialists, on the basis of their all-importance to employment and regional economic development, supposedly worthy in that sense of special protection—special privilege, as Tom L. Johnson now saw it. Suddenly, after Johnstown, the industrial corporation was just as often described in court cases, and throughout American culture, as too powerful, reckless, manipulative of the public good, deserving of no special protection at all. Sometimes big, awful events carried by news media cause change more quickly and definitively than scholarship or argument or protest ever can. The simple idea that those who create risk, and have the ability to reduce risk, must bear responsibility for costs of that risk—even when all they may have intended was restoration of their souls by healthful recreation—came to seem, for the first time, mainstream. Soon millionaires could no longer realistically presume that heedlessness would bring no dire consequences or loss down on their own heads. From that change, much would ensue.

The conflict, in the end, was between the people and the monopoly. In 1890, Congress passed the Sherman Antitrust Act. In 1901, President Teddy Roosevelt started using that act to break up monopolies. In the meantime, the labor organizing that men from Andrew Carnegie to Daniel Morrell had been so adept at suppressing strengthened beyond what even men with that degree of power could stop. The early years of the new century would see outright warfare between labor and capital in the embattled but ultimately successful American labor movement. In a host of areas, from factory conditions to housing to children's rights, reformers of all kinds would begin to find new ways of using government to inspect, regulate, fine, and otherwise control the otherwise limitless power of wealth to destroy lives.

The ground had shifted. As the twentieth century began, new conflicts opened that would, in their own ways, change the face of American society yet again. As the 1890s legal cases show, it was the Johnstown Flood that began that shift, changing not only American law but all of American culture.

SONG AND STORY

THE SOUTH FORK FISHING AND HUNTING CLUB SOON FOLDED. Not only was there no lake, but the club was nationally infamous now, associated with horrible destruction, and the cottages and boardwalks and clubhouse were hardly any place for relaxation and recreation.

The clubhouse did open for the summer of 1889, with the new indoor-plumbing system largely in place, but it had little appeal, and over time the cottages were simply abandoned. A caretaker lived in the clubhouse, but the place where amateur theatricals and band music and bass fishing had gone on during the summers of the 1880s became a kind of ghost resort.

In early 1904, the club's organization dissolved, too, with an auction sale of everything still in the clubhouse. The auction was held up on the mountain, at the clubhouse itself; people went up and bid on the contents of fifty bedroom suites, yards and yards of carpet, silverware engraved with the club monogram, all kinds of furniture. Shortly afterward, coal was discovered in the ground below the clubhouse, where the lake had been, and in

1907, the Maryland Coal Company sank a shaft there, put managers in some of the old cottages, and built new worker housing. For a long time the clubhouse operated as a hotel and bar.

By the time of the clubhouse auction, Andrew Carnegie and Henry Clay Frick had consolidated all of Carnegie's steel ventures into one mighty company. What had begun as the Edgar Thomson Works at Braddock, near Pittsburgh, had already spread to the nearby Homestead and Duquesne works. Now Frick's vision was to conglomerate those three mills with others they'd bought and form the Carnegie Steel Company, with a fifteen-story steel showplace of a headquarters in downtown Pittsburgh. That move created one of the biggest companies so far, in the monopoly approach to capitalism that was beginning to draw the concerned attention of reformers.

Even as Carnegie and Frick were making that move, a strike broke out, as they used to say, at the Homestead mill. Really, it didn't break out: that was how most strikes had occurred before, in sudden, ad hoc uprisings of infuriated workers. The Homestead Strike was especially disconcerting to Carnegie and Frick because it was planned. It was organized. It was led by the Amalgamated Association of Iron and Steel Workers, the craft union for the skilled laborers in the mills.

For years Frick had been hoping to break that organization. After an earlier strike, the union had essentially taken over and was running the works; Frick believed its rules were badly limiting productivity. Now he saw his chance.

Carnegie, busy traveling the world in advanced and high-minded circles, had proclaimed his support for labor unions, even sometimes called himself a socialist. But union rules at Homestead had been putting more men to work than he considered necessary. Cost-cutting was always Carnegie's chief concern, and abstract support for labor didn't mean that the union could be further tolerated at Homestead. That's why Carnegie had Frick.

In one of the most famous showdowns in American history, Frick made his move. He locked out the workers, hired scabs, cut off collective bargaining, and brought in the notorious Pinkerton Agency, a mercenary military and police force. War broke out in July 1892—literally, as the workers and Pinkertons exchanged gun and cannon fire in a bloody battle—and even with some of the Pinkertons in surrender, Frick declined to negotiate to end the fighting. He knew that if the battle went on, the governor would send in troops.

The governor did: six thousand troops came to Homestead. The anarchists and free-love advocates Alexander Berkman and Emma Goldman came too, hoping to assassinate Frick, and when Berkman made the attempt, Frick won the war. In the wake of the assassination attempt, support for the union and the strike, even among other labor leaders, dried up, and the Amalgamated Association soon became a nonentity at the Homestead works and weakened quickly at other shops as well.

So Carnegie Steel flourished, largely under Frick's management. Soon they'd captured, as Carnegie liked to put it, 25 percent of the nation's steel market. They sold the company in 1901, to the financier J. P. Morgan, and that was the biggest transaction, and the biggest industry consolidation, up to that point in history. Morgan's resulting company, U.S. Steel, combined Carnegie Steel with the Federal Steel and the National Steel companies in a buyout totaling $492 million—more than $14.02 billion in 2018 terms—with a capitalization of $1 billion and Frick's man Charles Schwab in charge. That first billion-dollar company became known on Wall Street simply as "the Corporation."

For U.S. Steel made no pretense of creativity, innovation, or even efficiency. The pitch to investors was hugeness, the sheer, overwhelming dominance of the Corporation, with total control of two-thirds of American steel production and an anticipation of possibly one day controlling it all. J. P. Morgan wasn't, putting it mildly, a steel man. He was a moneyman, and what a moneyman he was.

It didn't work out perfectly. Schwab went over to the competitor Bethlehem Steel—also intent on dominance—where things were more fun: the smaller conglomerate did have to innovate to compete. As early as 1911, U.S. Steel and Bethlehem Steel were pretty much splitting the American market. In the process, however, Andrew Carnegie, having gotten out of steel with a staggering payoff, became far richer than he'd ever been before. It was in the early twentieth century that he really started giving his money away and doing good.

One of the things Tom L. Johnson learned from the Johnstown Flood was that he disliked charity. Tom's frustration with the process of distributing all the extra money that had come into Johnstown was based on a frustration with certain aspects of human nature.

Nobody had worked harder to dig out the city than Tom, and nobody had a higher appreciation for those like Captain Bill Jones, who rolled up their sleeves and set to work to help. Tom was amazed and impressed by the speed of recovery that came from the massive group effort. He wasn't against help; he wasn't even against what he knew were the generous impulses that made people want to send money and supplies.

But Tom had been struck, during the Johnstown recovery, by what he saw as a weird human inconsistency. People complacently tolerated the daily horrors of poverty that were marking the end of the nineteenth century: child labor, the overworking and underpaying of women. Among the more prosperous classes, there didn't seem to be much sympathy for those sufferings. When a horrible situation was normal, people just accepted it and looked away. Yet in response to the unusual, to news—earthquakes, fires, kidnappings—sympathy exploded in a kind of maniacal flood of its own. When an event of that kind jumped from the headlines, the entire American public would empty its purses and wallets and smash its children's piggy banks to blanket the victims in financial mercy. Tom had seen that in the reaction to Johnstown.

But there would be no need for all this frenzied pity if there were no special privilege, no free rein for greed. That had become Tom's view. Charity could do nothing to restore children to parents, heal broken hearts, or bring back lost lives. The Johnstown disaster never would have happened in the first place if commonsense restraints had been placed on privilege. What really irritated Tom, as the city rebuilt, and as it seemed to become miraculously even stronger and better than before, was hearing people say that, in the end, "the flood was a good thing for Johnstown." People said that about the Chicago fire, too, and about other disasters from which cities made impressive recoveries.

But real people had died in Johnstown, by the thousands. Losses like that couldn't be recovered by material good. The way Tom saw it, the Johnstown Flood had made it literally true that special privilege destroys life and causes unutterable pain. Had there been a way to prevent the rich men on the mountain from making a lake on their own passing whim, or, if they had to make a lake, to require them to build and maintain a proper dam to hold it in, there would have been no need for charity at all.

So the failure, in the end, was with human government. Tom didn't hate charity itself. He hated the human preference for tinkering forever, as he put it, with a defective spigot when the bunghole is wide open. Finding the cause instead of responding to symptoms, using the collective power of democratically elected government to repair the cause and render charity unnecessary: that offered the only hope for the American future that Tom could see.

When he became the mayor of Cleveland in 1901, nationally famous as one of the first progressive mayors of a major American city, Tom L. Johnson began to do just that. He held office for four terms and changed everything about the way Cleveland operated. He brought his natural ebullience to bear: as mayor, Tom became famous for, among other things, striding into the city's most elegant public park, pulling up a "Do Not Walk on the Grass" sign, and tearing it up, to loud public acclaim. He kept his sense of humor and his idiosyncratic attitude, and when he was done governing, it was the municipality, not private franchises, that delivered essential services. Tom L. Johnson had gone from a dedicated monopolist to what might have seemed to some a wild-eyed socialist. But what he really did in Cleveland, directly inspired by the horror of the 1889 Johnstown Flood, was create a modern, working city.

In Johnstown, Tom's partner Arthur Moxham led the related fight to modernize the city by consolidating its fragmented government. Further terrible flooding would plague Johnstown. Yet the event of a huge private lake dropping down on the city, solely because of the lake's owners' disregard, had been eminently preventable. It hadn't been prevented, in the view of Moxham and Johnson and others, because efforts at prevention had been left to conversations among leading citizens: Daniel Morrell's personally expressing displeasure. Benjamin Ruff's blowing him off.

Blame in that sense lay, to Moxham, with Johnstown itself. The political citizenry had engaged in what he called "criminal irresponsibility." The facts had been known: a dangerous body of water existed, and nobody trusted the dam that held it back. Yet the whole community—as a community—hadn't even investigated the situation. No organization had existed, representative of the community as a whole, with the power to do so. Had there been such an entity, the South Fork dam would have been dealt with long since, and what happened to everybody in the valley on May 31 wouldn't have happened. It was that stark, that simple.

Strong words. Regulating big dams now seemed like a good idea, and without a government to do so, the regulation couldn't occur. In the very spring of the recovery, Johnstown was chartered as a city.

The Cambria company more than recovered, and the smoking, sparking, and roaring of its long brick buildings and high stacks only spread down the river. The valley's industrial might began amplifying to the extraordinary proportions that would make western Pennsylvania a major commercial and industrial center for much of the twentieth century. As the region not only recovered but kept booming, consolidation of big-scale industry came to Johnstown in 1916, when Cambria was bought by the Midvale Steel and Ordnance Company. In 1923, Midvale sold the works to Bethlehem. Charles Schwab, having begun under Carnegie, and having then gone to U.S. Steel in the buyout, had made Bethlehem the second-largest steel company in the world, and the company's famous Johnstown Works would operate until 1992.

◆ ◆ ◆

There was just something special about the story of the Johnstown Flood. Maybe the power it wielded for so many years over the American imagination came from its status as the biggest national story after the close of the Civil War. Maybe the protest against the heedless rich that the flood engendered served as a sign of new times. Maybe the story resonated with a widespread anxiety over big industry and the enormity of the things that could go wrong now. But whatever the factors, and there were probably many, the story of the Johnstown Flood quickly took on a life of its own that would forever set it apart, in American popular culture, from other disaster stories, even from worse ones to come.

Disaster tourism had always existed. After the flood, it exploded in Johnstown. Even while the citizens and the committees and the commissions were trying to clean up and provide relief, under enormous logistical and emotional pressure, trains were bringing in passenger cars full of tourists hoping to snatch up objects, or just to gawk at the astonishing misery and destruction that lay about them. These visitors were unwelcome, to say the least, but it was hard to keep them out.

Troops were stationed at the pontoon bridge, and later at the partially repaired stone bridge, to ensure that nobody came into town who didn't have business there. But hundreds of coffins were being sent from Pittsburgh for use by the Reverend Beale's morgues, and that gave the tourists an idea. They started carrying coffins, and they gained entry that way.

Having arrived, most didn't help. They watched, and they took things. One souvenir hunter saw a wooden leg sticking out of the pile and went after the trophy. As he and his friends dug and pulled at it, they found the leg was attached to a dead body. Others just stood around and gathered stories. There were so many moving sights to see, to take home and tell friends about, proof of having been right there in Johnstown.

Stories of devoted motherhood were big in 1880s America, and many victims of the flood, as their bodies were pulled from the wreckage, gave testimony to that devotion. Entire dead families were found together, and sometimes the mother seemed to be firmly holding a baby even in death. One still had her baby at the breast. Two sisters were found locked in an embrace. Some of these things were real, others exaggerated or invented, but the eager tourists, packing back onto the passenger cars to get out of there at the end of the day, didn't care. They'd been there. They'd seen it.

Not everybody with a desire to experience the Johnstown Flood could actually get there and see it. An entire industry quickly developed to satisfy their needs. It wasn't just the constant newspaper coverage, dredging up story after story. Whole books started coming out right away. One was titled *The Johnstown Horror!!! or Valley of Death: Being a Complete and Thrilling Account of the Awful Floods and Their Appalling Ruin, Containing Graphic Descriptions of the Terrible Rush of Waters; the Great Destruction of Houses, Factories, Churches, Towns* . . . (and it actually went on from there).

That was probably the most extreme, but it was a bestseller, and the idea behind all of these books was the same. Most came out only months after the flood, when the real city was still a wasteland and there had been no time to assess what actually happened.

Photographers too came to town and set up their big, boxy cameras and came away with troves of images that started showing up on 3-D viewers, postcards, and as illustrations for the books. Books that couldn't get photos had paintings and engravings. Traveling showmen projected flood images on walls—the "magic lantern"—and gave fanciful lectures on the event. Poetry was a highly popular medium then, and poets went at the flood with gusto. The *New York World* commissioned Walt Whitman, America's greatest poet of the day, to crank something out, and Whitman may have surprised his sponsor by writing, as usual, something highly counterintuitive:

> War, Death, cataclysm like this, America,
> Take Deep to thy proud, prosperous heart.

Many lesser lights wrote more accessible Johnstown poems. All that was just at first blush. With time, things really got ambitious. The big thing became Johnstown Flood reenactments, with a feeling, anyway, of audience participation. In 1901, at Coney Island, the world-famous bayside amusement park in Brooklyn, New York, exhibitors built a giant, turreted castle on Surf Avenue, above whose big, Gothic-arched doorway the words JOHNSTOWN FLOOD invited the curious inside. The show—it had come straight from the Buffalo, New York, world's fair, where it had been a sensation—featured a cyclorama, with scenes acted out in front of gigantic panoramic paintings: movies hadn't yet taken off, and this effect was riveting for both realism and spectacle. Real water poured, and electrical sound effects—explosions, booming, grinding—made the audience jump and scream. The show ran for a number of years. In her early twenties, Gertrude Quinn saw the Coney Island show on a trip to New York. The show didn't do much for Gertrude, not surprisingly, but it was such a big hit that in 1906 the impresarios upped the ante and decided to do the story of Noah's ark instead, with not Johnstown, Pennsylvania, but the whole world getting destroyed, and not just by flood but by earthquake and fire, too; why not? Meanwhile similar Johnstown shows went on at the White City exposition in Chicago, Atlantic City in New Jersey, and elsewhere.

These shows and books and paintings made use of thrilling stories that weren't true. A favorite and lasting one was the story of the Paul Revere of Johnstown. This fable had been concocted right after the flood itself, and always involved, in its various versions, a courageous young man who galloped his horse down the valley immediately ahead of the great wave, calling out a warning and thus saving thousands of lives before being tragically overtaken by the flood and drowned. That this act would have been physically impossible made no difference. It was too good to be true, but too good not to believe.

Another good one was drawn from the death of Hetty Ogle, the telegraph operator who had stayed at her post until the office got too deep in the water. Mrs. Ogle had finally left the office and gone up to the third floor. She died there, with her daughter, when the wave hit town. But before she'd left she'd tapped out "This is my last message," and while she'd meant only that it was her last of the day, in the story of Hetty Ogle, heroine, she'd been tapping out warnings and saving lives even as the wave was rampaging through town. And she'd known that would be her last message ever.

Meanwhile, a trove of songs were placing the Johnstown Flood in the American pop archive. These songs were sold as sheet music, to be sung at home to the accompaniment of a parlor piano, but they were also passed around and changed and embellished by musicians learning them by ear and playing along on guitars, jugs, and other less expensive instruments. The "sentimental song" was a genre, meaning a song invoking strong feelings, and so was the "event song," usually describing some recent disaster. Johnstown was fodder for both, and songs with titles like "That Valley of Tears" and "My Last Message" were popular. A song titled simply "The Johnstown Flood" made the Paul Revere of Johnstown its hero.

The most mysterious of the flood songs may be "The Night of the Johnstown Flood," mentioned in the 1982 song ("Highway Patrolman") by Bruce Springsteen. While no song with that title had existed when the Springsteen song was released, people inspired by its mention have written some.

Victor Heiser became a doctor. Having first hoped to be a watchmaker, he worked as a carpenter and a plumber and ended up attending Jefferson Medical College in Philadelphia. Fluent in four languages, for a time he worked checking immigrants for infectious diseases at Ellis Island, and when American military forces took over the Philippines in 1898 he was appointed director of health for the island group. Victor oversaw what was known as the Culion Leper Colony: "the island of no return," with at times five thousand patients and two hundred doctors.

After being relieved of duty there, between 1903 and 1915, he oversaw the creation of the entire public health system for the American Philippines, with a major focus on treating leprosy. After that, he worked for the Rockefeller Foundation, traveling the world promoting public health and fighting not only leprosy but also smallpox, plague, cholera, malaria, and beriberi.

Early in his career, Victor Heiser had set a personal goal. He would save fifty thousand lives per year from preventable diseases. He died in 1972. People said that in the end, he might have saved more than two million.

Like Victor Heiser, the Reverend David Beale left Johnstown not long after the flood. During the grueling recovery effort, he'd volunteered his Presbyterian church for use as a morgue, and the church board, offended, scolded him for having failed to ask permission. Beale, his back up, argued that it had seemed the obvious and the Christian thing to do, but there was no compromise. He left town to serve elsewhere.

The great steel man Captain Bill Jones arrived home in Pittsburgh after his two straight weeks of hard work in the Johnstown recovery and went back to work at Carnegie's Edgar Thomson Works. In late September of that year, he was inspecting a faulty furnace when it broke, pouring down hot coal and metal. Bill, badly burned, fell and struck his head. He died at the Homeopathic Hospital in Pittsburgh. His funeral was attended not only by Andrew Carnegie and Henry Frick and steel executives of every other company but also by about ten thousand steelworkers.

Thanks to its amazingly successful efforts in Johnstown, the Red Cross became nationally known as the leading relief organization in the country. Clara Barton would go on to bring massive relief to the Russian famine, the Sea Islands hurricane, and the Armenian massacres. Her final trip, taken when she was seventy-eight, was in relief of the destruction by hurricane of the city of Galveston, Texas, in 1900, which she would say dwarfed anything she'd seen before, even Johnstown: Galveston would remain the worst natural disaster in American history—though it had its shameful human element, too—for at least the ensuing 117 years. At the age of eighty-three, Barton was forced to give up leadership of the Red Cross, and she died in 1912.

General Hastings, for his part, gobsmacked by yet ever courteous to Miss Barton throughout the whole Johnstown endeavor, ran for governor of Pennsylvania in 1890 but failed to get the Republican nomination. In 1894 he did get it, defeated the Democratic nominee, and served one term.

◆ ◆ ◆

Strangely enough, Gertrude Quinn and Max McCachren never saw one another again after her rescue. Class distinctions were rigid: nobody on either side would necessarily have thought it appropriate for the little girl and the big house painter who had carried her downstream that crazy night, and thrown her through the air, to have a reunion.

They thought of one another often. Max liked to drop by the Quinns' new store, and he and James and Rosina Quinn would relive the rescue: James would often slip Max five dollars from the till in a celebration of the event. Gertrude did see the Metz family that had taken her in—she often happily carried gifts from the store to their house—but as an adult she came to consider it one of the great regrets of her life that she'd never seen Max again.

She was living in another town, Gertrude Slattery now, with children of her own, when she saw his obituary notice in the *Johnstown Tribune*. She sent dark red roses. Gertrude's younger sister Eulalia, living in Johnstown, went to the funeral home to pay respects: the family had placed the roses on his casket. Max's family sent Gertrude his picture, and she never parted from it.

Back when Max was making his regular visits to the new Quinn store, Papa came home one day and told Gertrude—then in her teens—that Max had once again been going over that night on the raft, and he'd told James of a great feeling of joy and love that had filled his heart, as the little girl had put both arms around his neck and held on to him for dear life.

"You may tell Mr. McCachren for me," Gertrude told Papa, "the next time you see him, that I said: 'When I'm grown up and meet the man I love, and put my arms around his neck, I am sure his neck will never feel as fine to my arms as the burly, unshaven neck of Maxwell McCachren, the brave Scotchman who risked his life to save me from the greatest flood since Noah's'!"

ACKNOWLEDGMENTS

Thanks to the Johnstown Flood Museum, the Johnstown Area Heritage Association, the Cambria County Library, and the Johnstown Flood National Historic Site.

A NOTE ON SOURCES AND FURTHER READING

For many years, the 1889 Johnstown Flood was one of American entertainment's most popular subjects. A wealth of primary and secondary material proliferated for almost a century after the disaster, from eyewitness accounts to reporters' narratives, from photographs to movies and songs. In confronting the nature and quantity of that material, *Ruthless Tide* is indebted to modern historical and scientific works that place the earlier record in responsible perspective.

The best-known modern book on the subject is David McCullough's groundbreaking *The Johnstown Flood,* first published in 1968; its nearest predecessor similar in scope was Richard O'Connor's *Johnstown: The Day the Dam Broke,* published in 1957. O'Connor's book mentions interviews with eyewitnesses but gives no specifics. McCullough's book, by contrast, remains notable in part for reliance on detailed, recorded interviews with survivors, Victor Heiser most comprehensively. In the half century since the book was published, some have dissented from aspects of its interpretation; some have emphasized other aspects of the primary record, as this book does as well. None, however, has superseded McCullough's work, and many who write on the flood itself, and on related matters like the life of Henry Clay Frick, the development of the steel industry, and the history of western Pennsylvania, have followed it closely.

McCullough has served *Ruthless Tide* not only as a source— Chapter Fourteen follows him especially closely on post-flood publicity and the failed efforts to litigate—but also as an introduction to the primary and eyewitness records on which this book is largely based. Another major modern secondary source used here to similar ends is Michael R. McGough's highly detailed *The 1889 Flood in Johnstown, Pennsylvania* (2002). Despite McCullough's and McGough's occasional conflicts, or perhaps because of them, reading the two works in concert offers a more comprehensive view of the flood, its background and ramifications, and the primary record than was previously available.

Other important secondary work recently available and relied on here: the 2016 hydrological study by Coleman, Kaktins, and Wojno; Steven Ward's 2011 physics-based flood simulation; Emily Godbey's 2006 study of disaster tourism at Johnstown; the dispute on the recovery effort carried on in the 1990s in *Pennsylvania History*; and Jed Shugerman's legal analysis of the advent in the courts of "strict liability" (2000). Going back to 1940, Nathan Shappee's unpublished doctoral dissertation remains the only full-scale scholarly study of the early development of Johnstown, its sudden industrialization, and the ecological heedlessness that made regular spring flooding so threatening even before the dam broke.

The primary record from which McCullough, McGough, and other writers have drawn their narrative accounts is rich. Some of the survivors were able writers of books: The most comprehensive and responsible overall contemporaneous account is Rev. Beale's *Through the Johnstown Flood,* packed with detail on the politics of recovery, sanitation, and emergency government. J. J. Mclaurin's *The Story of Johnstown,* published, like Beale's book, only a year after the event, has long served modern writers as a source of flood stories. Clara Barton sets out her point of view on General Hastings and the relief and recovery effort in general, as followed closely here, in the Johnstown chapter of her book *A Story of the Red Cross*.

With the comparatively recent development of the Internet, much of the primary record has become directly available not only to researchers and authors but also to the general public. The Johnstown Flood Museum, the Johnstown Area Heritage Association, and the Johnstown National Memorial have put many of those sources online; the memorial has also posted nearly the entire body of the important testimony given during the railroad company's investigation of the disaster.

Eyewitnesses of course disagree, and sometimes even with themselves. John Parke's two accounts of the day the dam broke—one in the Special Report of June 1889 and one in the Report of the Committee of 1891—differ in some particulars, causing discrepancies in later accounts. Club members contradicted themselves. Testimony of railroad company employees could be self-serving and confusing. Like earlier books, *Ruthless Tide* seeks to weave existing strands into a cogent narrative; in the bibliography, interested readers will find many paths for exploring texture and controversy in the underlying record.

The narrative and interpretive approaches taken in *Ruthless Tide* rely to an unusual degree on two first-person accounts that other books have cited only lightly: Gertrude Quinn Slattery's *Johnstown and Its Flood* and Tom L. Johnson's *My Story*. Each in its own way is fresh and idiosyncratic; readers of *Ruthless Tide* who get interested in Gertrude and Tom are likely to enjoy their books, too. On the larger matters of monopoly power, the iron and steel industries, the rise of the Pennsylvania Railroad, and the nascent labor movement, readers may especially enjoy, for example, David Nasaw on Carnegie, David Brody on pre-union steelworkers, and Albert Churella on the railroad.

As to location scouting, the two museums and the national

memorial mentioned above as online resources fully repay visiting in real life. Related sites in southwestern Pennsylvania very much worth the trip include the Staple Bend Tunnel, the Allegheny Portage Railroad National Historic Site, the Railroaders Memorial Museum, and the Horseshoe Curve National Historic Landmark.

All works mentioned in this note, along with the others behind the story told in *Ruthless Tide,* are cited in full in the bibliography.

BIBLIOGRAPHY

"Andrew Carnegie: The Richest Man in the World." *American Experience.* http://www.pbs.org/wgbh/americanexperience/films /Carnegie/.

Barton, Clara. *A Story of the Red Cross: Glimpses of Field Work.* New York: D. Appleton and Co., 1917.

Beale, David J. *Through the Johnstown Flood.* Philadelphia, PA: Edgewood Publishing Company, 1890.

"Before the Reservoir Came: Editor George Swank, June 14, 1889." Johnstown Flood Museum. http://www.jaha.org/edu/flood/story /swank_trib_14jun_01.html.

"Benjamin Franklin Ruff (1835–1887)." Johnstown Flood National Memorial. https://www.nps.gov/jofl/learn/historyculture /benjamin-franklin-ruff.htm.

Bremner, Robert H. "The Civic Revival in Ohio: Reformed Businessman: Tom L. Johnson." *The American Journal of Economics and Sociology* 8, no. 3 (April 1949).

Brody, David. *Steelworkers in America: The Nonunion Era.* Urbana, IL: University of Illinois Press, 1998.

"Canals in Pennsylvania." North East Rails. http://www.northeast .railfan.net/canal.html.

"Captain William R. (Billy) Jones." The Hopkin Thomas Project. http:// himedo.net/TheHopkinThomasProject/TimeLine/Genealogy Portraits/CaptWRJonesBios/CaptWilliamRJones.htm.

Carnegie, Andrew. *The Autobiography of Andrew Carnegie.* New York: Public Affairs, 2011.

————. *The Gospel of Wealth, and Other Timely Essays*. Cambridge, MA: Belknap Press of Harvard University Press, 1962.

Casson, Herbert N. *The Romance of Steel: The Story of a Thousand Millionaires*. New York: A. S. Barnes & Company, 1907.

Churella, Albert J. *The Pennsylvania Railroad, Vol. 1: Building an Empire, 1846-1917*. Philadelphia: University of Pennsylvania Press, 2012.

Coleman, Neil M., Uldis Kaktins, and Stephanie Wojno. "Dam-Breach Hydrology of the Johnstown Flood of 1889: Challenging the Findings of the 1891 Investigation Report." *Heliyon* 2 (2016). http://www.heliyon.com/article/e00120/.

"Colonel Elias J. Unger (1830–1896)." Johnstown Flood National Memorial. https://www.nps.gov/jofl/learn/historyculture/colonel-elias-j-unger.htm.

"The Conemaugh Disaster: How a Repetition May Be Avoided." *The Colliery Engineer* 10, no. 1 (August 1889).

"Cresson Resort Was First Club." *Altoona Mirror*, July 24, 2015. http://content.altoonamirror.com/?p=616240/Cresson-resort-was-first-club.html.

Dietrich, William S. "Henry Clay Frick: Blood Pact." *Pittsburgh Quarterly* (Spring 2009) https://pittsburghquarterly.com/pq-people-opinion/pq-history/item/329-henry-clay-frick-blood-pact.html.

"Flood of 1889." Histories of the National Mall. http://mallhistory.org/items/show/453.

"Gilded/Progressive Age—Then." Gilded Age Photography. https://gildedagecuration.weebly.com/gildedprogressive-age---then.html.

Godbey, Emily. "Disaster Tourism and the Melodrama of Authenticity: Revisiting the 1889 Johnstown Flood." *Pennsylvania History: A Journal of Mid-Atlantic Studies* 73, no. 3 (Summer 2006).

"The Great Storm of 1889." Johnstown Flood Museum. http://www.jaha.org/edu/flood/why/rain_greatstorm.html.

Hartman, Jesse L. "The Portage Railroad National Historic Site and the Johnstown Flood Memorial." *Pennsylvania History: A Journal of Mid-Atlantic Studies* 31, no. 2 (April 1964).

"Henry Clay Frick." The Frick Collection. https://www.frick.org /about/history/henry_clay_frick.

Henshall, James A. *Book of the Black Bass, Comprising Its Complete Scientific and Life History* . . . Cincinnati, OH: R. Clarke and Co., 1881.

Historic Structures Report, Architectural and Historical Data Section: Clubhouse, Brown Cottage, Moorhead Cottage, and Clubhouse Annex—South Fork Fishing and Hunting Club. Southwestern Pennsylvania Heritage Preservation Committee and South Fork Fishing and Hunting Club Historical Preservation Society, 1993.

Historic Resource Study: Cambria Iron Company. America's Industrial Heritage Project, Pennsylvania. United States Department of the Interior/National Park Service, 1989.

"History of Coal in Cambria County." Johnstown Area Heritage Association. http://www.jaha.org/attractions/heritage-discovery -center/Johnstown-history/history-coal-cambria-county/.

"History of Steel in Johnstown." Johnstown Area Heritage Association. http://www.jaha.org/attractions/heritage-discovery-center /Johnstown-history/history-steel-Johnstown/.

Hogeland, William. *The Whiskey Rebellion: George Washington, Alexander Hamilton, and the Frontier Rebels Who Challenged America's Newfound Sovereignty.* New York: Simon & Schuster, 2010.

"Iron Meteorites: The Hearts of Long-Vanished Asteroids." Geology .com. http://geology.com/meteorites/iron-meteorites.shtml.

Johnson, Tom L. *My Story.* Edited by Elizabeth J. Hauser. New York: B. W. Heubsch, 1911.

Johnson, Willis Fletcher. *History of the Johnstown Flood, Including All the Fearful Record.* Philadelphia, PA: Edgewood Publishing Company, 1889.

"The Johnstown Flood Show (1902–1905) and the Deluge (1906–

1908)." Luna Park: Heart of Coney Island. http://www.heartof coneyisland.com/Johnstown-flood-show.html.

"Johnstown [Steel] Historical Marker." ExplorePAhistory.com. http://explorepahistory.com/hmarker.php?markerId=1-A-23A.

"Johnstown's Immigration History." Johnstown Area Heritage Association. http://www.jaha.org/attractions/heritage-discovery -center/Johnstown-history/Johnstowns-immigration-history/.

Kraussjune, Margaret J. "Why Does Pennsylvania Have Only a Handful of Natural Lakes?" *The Allegheny Front*, June 16, 2017. https://www.alleghenyfront.org/why-does-pennsylvania-have -only-a-handful-of-natural-lakes/.

McCullough, David G. *The Johnstown Flood*. New York: Simon & Schuster, 1968.

McGough, Michael R. *The 1889 Flood in Johnstown, Pennsylvania*. Gettysburg, PA: Thomas Publications, 2002.

McLaurin, J. J. *The Story of Johnstown: Its Early Settlement, Rise and Progress, Industrial Growth, and Appalling Flood on May 31st, 1889*. Harrisburg, PA: James M. Place, 1890.

McMaster, John Bach. "The Johnstown Flood: II." *The Pennsylvania Magazine of History and Biography* 57, no. 4 (1933).

McMaster, John Bach, and Ellis Paxson Oberholtzer. "The Johnstown Flood." *The Pennsylvania Magazine of History and Biography* 57, no. 3 (1933).

Miner, Curtis, and Richard Burkert. "The Myth of Demythification: A Response to Don Mitchell's 'Heritage, Landscape and the Production of Community.'" *Pennsylvania History: A Journal of Mid-Atlantic Studies* 59, no. 3 (July 1992).

Mitchell, Don. "Whose History and History for Whom? Questions About the Politics of Heritage: A Reply to Miner and Burkert." *Pennsylvania History: A Journal of Mid-Atlantic Studies* 59, no. 3 (July 1992).

Morawska, Ewa. *Insecure Prosperity: Small-Town Jews in Industrial America, 1890–1940*. Princeton, NJ: Princeton University Press, 1997.

Nasaw, David. *Andrew Carnegie.* New York: Penguin Books, 2007.

"National Park Service to Acquire South Fork Club, Which Was at Heart of Johnstown Disaster." *Pittsburgh Post-Gazette,* July 18, 2006.

O'Connor, Richard. *Johnstown: The Day the Dam Broke.* New York: J. B. Lippincott Company, 1957.

"The Paul Revere of Johnstown: True or False?" Johnstown Flood Museum. http://www.jaha.org/edu/flood/why/warning_system-paulrevere.html.

"The Pennsylvania Iron Industry: Furnace and Forge of America." ExplorePAHistory.com. http://explorepahistory.com/story.php?storyId=1-9-17&chapter=1.

"People: The South Fork Fishing and Hunting Club Members." Johnstown Flood National Memorial. https://www.nps.gov/jofl/learn/historyculture/people.htm.

"Popular Music Inspired by the Johnstown Flood." Johnstown Flood Museum. http://www.jaha.org/edu/flood/story/img/music/index.html.

"Profiles in Time: Meet the Members of the Club." http://profilesintime.blogspot.com/.

"Report of the Committee on the Failure of the South Fork Dam." *Transactions of the American Society of Civil Engineers* 24 (January–June, 1891).

Roker, Al. *The Storm of the Century.* New York: William Morrow, 2014.

Rosenthal, Ellen M. "Review of *The Johnstown Flood: The True Story of One of the Most Devastating Disasters in American History* by Charles Guggenheim." *The Journal of American History* 80, no. 1 (1993).

Shappee, Nathan Daniel. "A History of Johnstown and the Great Flood of 1889: A Study of Disaster and Rehabilitation." PhD diss., University of Pittsburgh, 1940.

Shugerman, Jed Handelsman. "The Floodgates of Strict Liability: Bursting Reservoirs and the Adoption of Fletcher v. Rylands in the Gilded Age." *Yale Law Journal* 110, no. 2 (November 2000).

Skrabec, Quentin R., Jr. *Henry Clay Frick: The Life of the Perfect Capitalist.* Jefferson, NC: McFarland, 2010.

Slattery, Gertrude Quinn. *Johnstown and Its Flood.* G. Q. Slattery, 1936.

"South Fork Dam." Johnstown Flood National Memorial. https://www.nps.gov/jofl/learn/historyculture/south-fork-dam.htm.

"South Fork Fishing and Hunting Club Repairs." Johnstown Flood Museum. http://www.jaha.org/edu/flood/why/dam_club_era.html.

"Special Report on the Johnstown Flood, June 1889." Johnstown Flood Museum. http://www.jaha.org/edu/flood/why/engineer_report.html.

Springsteen, Bruce. "Highway Patrolman." *Nebraska.* Columbia Records, 1982.

"Statement of A. H. Butler." Statements of Employees of the Pennsylvania Railroad Company. Johnstown Flood National Memorial. https://www.nps.gov/jofl/learn/historyculture/butler.htm.

"Statement of C. P. Dougherty." Statements of Employees of the Pennsylvania Railroad Company. Johnstown Flood National Memorial. https://www.nps.gov/jofl/learn/historyculture/dougherty.htm.

"Statement of Charles J. Moore." Statements of Employees of the Pennsylvania Railroad Company. Johnstown Flood National Memorial. https://www.nps.gov/jofl/learn/historyculture/moore.htm.

"Statement of D. T. Brady." Statements of Employees of the Pennsylvania Railroad Company. Johnstown Flood National Memorial. https://www.nps.gov/jofl/learn/historyculture/brady.htm.

"Statement of Emma Ehrenfeld." Statements of Employees of the Pennsylvania Railroad Company. Johnstown Flood National

Memorial. https://www.nps.gov/jofl/learn/historyculture/emma .htm.

"Statement of F. S. Deckert." Statements of Employees of the Pennsylvania Railroad Company. Johnstown Flood National Memorial. https://www.nps.gov/jofl/learn/historyculture/deckert.htm.

"Statement of H. M. Bennett." Statements of Employees of the Pennsylvania Railroad Company. Johnstown Flood National Memorial. https://www.nps.gov/jofl/learn/historyculture/bennett.htm.

"Statement of J. B. Plummer." Statements of Employees of the Pennsylvania Railroad Company. Johnstown Flood National Memorial. https://www.nps.gov/jofl/learn/historyculture/plummer.htm.

"Statement of J. C. Hess." Statements of Employees of the Pennsylvania Railroad Company. Johnstown Flood National Memorial. https://www.nps.gov/jofl/learn/historyculture/hess.htm.

"Statement of P. N. Pickerell." Statements of Employees of the Pennsylvania Railroad Company. Johnstown Flood National Memorial. https://www.nps.gov/jofl/learn/historyculture/pickerell.htm.

"Statement of R. C. Liggett." Statements of Employees of the Pennsylvania Railroad Company. Johnstown Flood National Memorial. https://www.nps.gov/jofl/learn/historyculture/liggett.htm.

"Statement of Robert Pitcairn." Statements of Employees of the Pennsylvania Railroad Company. Johnstown Flood National Memorial. https://www.nps.gov/jofl/learn/historyculture/pitcairn.htm.

"Statement of S. W. Keltz." Statements of Employees of the Pennsylvania Railroad Company. Johnstown Flood National Memorial. https://www.nps.gov/jofl/learn/historyculture/keltz.htm.

"Steel and Coal Company Towns." Johnstown Heritage Discovery Center. http://www.jaha.org/edu/discovery_center/work/img/company_towns/pages/cambria_smokeless_mine_c.html.

"Stereo (3-D) Views: Destruction Caused by the Flood." Johnstown Flood Museum. http://www.jaha.org/edu/flood/story/img/stereo-destruction/.

Stone, Katherine. "Origins of the Job Structure in the Steel Industry." Libcom.org. https://libcom.org/history/origins-job-structure-steel-industry.

Walker, James Herbert. *The Johnstown Horror, or Valley of Death*. Globe Bible Publishing Co., 1889.

Ward, Steven N. "The 1889 Johnstown, Pennsylvania Flood: A Physics-Based Simulation." In *The Tsunami Threat: Research and Technology*, edited by Nils-Axel Mörner, 447–66. InTech, 2011.

Winkelstein, Warren, Jr. "The Johnstown Flood: An Unnatural Disaster." *Epidemiology* 19, no. 1 (2008).

Yang, Heloisa, Matt Haynes, Stephen Winzenread, and Kevin Okada. "The History of Dams." University of California, Davis, Center for Watershed Studies. https://watershed.ucdavis.edu/shed/lund/dams/Dam_History_Page/History.htm.

INDEX

INDEX

Crawford Coke and Coal, 86
Cresson Springs, 37, 38–39, 63–64, 66, 74
Crime, 217–21
Cronyism, 44, 96
Culion Leper Colony, 274

D

Daisy (cow), 7
Darby, Abraham, 34
Deckert, Frank, 122, 123, 127–28, 199
Decoration Day (May 30, 1889), 101–2, 104–5
Deforestation, 13, 14, 15
Dick, Charlie, 106, 167–68, 169, 219
Dickens, Charles, 69
Disaster souvenirs, 218, 270–71
Disaster tourism, 270–74
Disease prevention, 221–22
Disinfection, 221–22
Donations, 230–33
Double-entry bookkeeping, 30
Dry-fly casting, 67
Dunfermline, Scotland, 26–27
Du Pont brothers, 96

E

East Conemaugh, 117, 121–22, 139–41
Edgar Thomson Steel Company, 59–60, 264, 275
Ehrenfeld, Emma, 119–22, 133–35, 199
Eiffel Tower (Paris), 197
Elk Lake, 74
Ellis Island, New York, 274
Epidemic, cholera, 71
Erosion, 14–15, 94
European immigrants, 55–56
Event songs, 273–74
Exposition Universelle (1889), 197–98

F

Fault doctrine, 257–58
Federal Steel Company, 265

Fenn, Mrs., 183
Financial Panic of 1873, 58, 76
Fires, 168–70, 212–13
First National Bank of Johnstown, 47, 85
Fishguard, 80, 112–13, 125–26, 250
Fishing, 67–68, 80, 92–93
Fishing lures, 67–68
Flood songs, 273–74
Fly fishing, 67
Fly rods, 67
Fort Pitt, 43
Foster, Barbara, 190, 192
Fourteenth Infantry Regiment, 221
Fourth Ward School, 209, 230
Franco-Prussian War, 180, 225–26
Frankstown Road, 179
Frick, Elizabeth Overholt, 75–76
Frick, Henry Clay, 75–77
    background of, 75–76
    Carnegie and steel industry, 76, 86, 91, 264–65
    Homestead Steel Strike and, 264–65
    post-flood response of, 198, 206, 253
    South Fork Club and, 75, 76–77, 84, 86
Frick, John, 76
Frick Coke Company, 76, 86, 91
Fritz, George, 7, 50–51
Fritz, John, 7, 50–51
Fulton, John, 210, 251
    dam inspection, 86–88, 89, 103, 248
Fulton, Mrs., 149

G

Galveston Hurricane of 1900, 12, 275
Garfield, James, 226
Gautier Wire Company, 3, 51, 143–44
Geis, Abbie, 9–10, 11, 13, 145–46, 155–57, 160–61, 238
Geis, Billy, 5

295

INDEX

INDEX

INDEX

New Austria, 143
New Florence Station, 203–5, 248
Newport, Rhode Island, 62
New York Sun, 250
New York World, 272
New York Yacht Club, 68
Niagara Falls, 181
Nickel, 32–33
"Night of the Johnstown Flood, The"
    (song), 274
Nitric acid, 222
Noah's Ark, 272
Nursing and Barton, 225

O
Ogle, Hetty, 127–28, 129, 273
Ohio River, 43, 44, 46
Oil, 29, 32, 35
Oil fires, 168
Oklahoma house, 242
Old Overholt, 75–76

P
Panic of 1873, 58, 76
Paris Exposition of 1889, 197–98
Parke, John, 102–3
    aftermath of flood, 249
    background of, 102
    dam breaks, 131–33
    flooding, 125–27, 128, 131
    inspecting the dam, 109–10
    rainfall, 105, 108–10, 111–13
    warning messages, 118–21
    work at club, 102–3, 104
Pennsylvania Main Line Canal, 69–70,
    72–73
Pennsylvania Railroad, 46, 103
    canal system and, 46, 72–73
    Carnegie and, 30–32, 35–36, 61
    flooding, 110–11, 199–205
    Great Railroad Strike of 1877, 57
    Horseshoe Curve, 70, 72

Pitcairn and, 84, 85, 91,
    199–205
Pennsylvania Supreme Court, 259
Phenique, 222
Phenyle, 222
Philippines, 274
Phipps, Henry, 91–92
Phipps, Henry, Jr., 206
Pig iron, 33, 49, 52–53
Pinkerton Agency, 265
Pitcairn, Robert
    background of, 91
    dam breaks and flooding, 199–203
    Pennsylvania Railroad and, 84, 85,
        91, 199–205
    relief efforts, 203–4, 206, 212
    South Fork dam and, 200–201,
        203–5, 248–49, 253
Pittsburgh
    Carnegie family move to, 27–28
    history of, 43–44
    relief train, 211–12, 213–14
    wealth of, 22
Pittsburgh Chamber of Commerce, 230
Pittsburgh Chronicle-Telegraph, 204
Pittsburgh Commercial Gazette, 204
Pittsburgh Dispatch, 204, 254
Pittsburgh Ladies' Committee, 213
Pittsburgh Locomotive Works, 34
Pittsburgh Relief Committee, 206, 210,
    211–12, 220, 229
Pittsburgh Times, 204
Placid, Lake, 63, 73–74
Poetry, 272
Policing, 217–21, 224–25
Pollution, 221–22
Poplar Street bridge, 124
Portage system, 69
Poverty, 97, 267
Presbyterian Church, Johnstown, 147,
    213, 274–75
Price-fixing, 61–62

INDEX

Progress and Poverty (George),
97–98
Prospect, 177
Psalm 23, 148
Psalm 46, 148
Purity, 222

Q

Quicklime, 222
Quinn, Eulalia, 276
Quinn, Gertrude, 4–11, 244
   aftermath of flood, 236–41
   background of, 4–5, 7–8
   dam breaks and flooding, 1–2,
      9–10, 145–46, 155–57, 160–61,
      173–75
   the fires, 167–75
   life after the flood, 272, 276–77
   newspaper reports of, 236
   rainfall begins, 106
   rescue of, 175, 185–90, 192–93,
      236, 276
   return to Johnstown, 241–43
Quinn, Helen, 156, 157, 191,
   192–93, 237
Quinn, James, 4–11
   aftermath of flood, 191–93,
      236–39, 242, 243
   appearance of, 5, 6
   concerns about dam safety, 94
   dam breaks and flooding, 9–11,
      16, 145–46, 156–58, 173–75
   life after the flood, 276–77
   rainfall begins, 106
   reputation for worry, 10–11
   rescue of Gertrude and,
      192–93
Quinn, Marie, 10, 242–43
Quinn, Rosemary, 156, 157, 191,
   192–93, 237–41
Quinn, Rosina
   aftermath of flood, 237–43
   background of, 5–6

in Kansas, 9, 237, 238–41
life after the flood, 276
Quinn, Vincent, 6–7, 8
   dam breaks and flooding, 9–10,
      146, 155, 158–59
   death of, 158–59, 238, 240
   rainfall begins, 106
Quinn's Store, 4, 47, 190, 191, 237,
   238, 276

R

Radioactivity, 32
Rainfall, 99–113
Red Cross, 225–30, 241, 275
Red Cross hotels, 229, 241
Reed, James Hay, 92, 254–55
Regulations, 28, 85, 269
Regulator Rebellion, 43
Reilly, John, 74–75, 200
Relief train, 211–12, 213–14
Republican Party, 8
"Retort" coke ovens, 47–48
Revere, Paul, 273, 274
Reverend Beale. *See* Beale, David
Revolutionary War, 43, 45
Robert the Bruce, 29
Rockefeller Foundation, 274
Roebling, John, 42, 46
Roosevelt, Theodore, 260
Rosin, 222
Ruff, Benjamin Franklin, 73–81
   background of, 65–66
   controlling interest in South Fork
      Club, 75, 76–77
   death of, 95
   desire to create a lake, 65–66,
      68–69, 73–74
   Morrell's concerns about dam,
      88–90, 95
   purchase of property for South Fork
      Club, 74–75
   redamming of South Fork, 77–81,
      83–84, 86–88, 200–201, 248, 250

INDEX

AL ROKER is cohost of NBC's *Today*. He has received thirteen Emmy Awards, ten for his work on *Today*. He is the author of *The Storm of the Century,* an acclaimed history of the 1900 Galveston hurricane. He lives in Manhattan with his wife, ABC News and *20/20* correspondent Deborah Roberts, and has two daughters and a son.